FUNDAMENTALS OF POWER INTEGRITY FOR COMPUTER PLATFORMS AND SYSTEMS

FUNDAMENTALS OF POWER INTEGRITY FOR COMPUTER PLATFORMS AND SYSTEMS

J. TED DIBENE II

WILEY

Published by John Wiley & Sons, Inc., Hoboken, New Jersey
Published simultaneously in Canada

Library of Congress Cataloging-in-Publication Data:

DiBene, J. Ted., II (Joseph Ted)
Fundamentals of power integrity for computer platforms and systems / J. Ted DiBene II.
pages cm
Includes bibliographical references and index.
ISBN 978-1-118-09143-2 (cloth)
1. Electric power system stability. 2. Electric power systems–Quality control. I. Title.
TK1010.D53 2014
621.31–dc23

2013030215

10 9 8 7 6 5 4 3 2 1

To my wife,
for her strength, love,
and courage

CONTENTS

FOREWORD

The development of computing hardware operating at increasingly higher speeds and requiring more power continues at an inexorable pace. Successful development of computing systems requires careful design of hardware so that unintentional analog effects do not seriously compromise or degrade digital performance. This is particularly true with systems operating at high clock speeds and having high power requirements. There are three design arenas that are crucial to successful digital operation of hardware: signal integrity (designing to ensure sufficient integrity of the signal waveform) power integrity (designing to maintain sufficient quality of the power supplied to active devices), and electromagnetic interference/compatibility (EMI/EMC) where the design is tailored to ensure that radio frequency emissions from the digital system do not violate international regulatory limits that are in place to protect the public airwaves. Typically, there are substantial areas of overlap in these design disciplines and in the specific design of any digital hardware system. Each discipline has been studied at length, but the push of faster and higher power hardware requires continued development of the design technologies and techniques embodied in each arena.

This book primarily addresses power integrity and offers an introductory-to-intermediate view of the requirements and design ramifications based on physics fundamentals, rather than on detailed mathematical modeling. A value in this book is that it provides the basic information to allow a problem to be defined without the need for creating a complex mathematical model and also provide means of checking the reasonableness of results obtained from complex models. Power integrity has typically been addressed in the literature as a subtopic of signal integrity at the printed circuit board level, so the author's system view and the consideration of the power integrity of both the integrated circuit package and the integrated circuit die is a valuable contribution to this field and should provide interesting reading for those pursuing this topic. The author has considerable practical experience in power and signal integrity design in the semiconductor industry, which lends credence to this book. I recommend this book to the reader and wish the author much success with its publication.

JAMES L. KNIGHTEN, PHD

Hardware Engineer at Teradata Corporation
IEEE Fellow
San Diego, CA
July 25, 2013

PREFACE

This book is an introductory text on power integrity. It is intended for students at the college under graduate level and for engineers who are new to the area of power integrity. It is assumed that the reader has some background in electromagnetics and basic power conversion. However, it has been written with an understanding that many concepts may be foreign even to engineers and thus the basics are covered first. It is also assumed that the reader has a working knowledge of how to use various tools, such as SPICE and math programs, for analysis. This text is *not* intended to teach modeling methods and how to use various field solvers—there are many good texts on these subjects that can be easily found through a search on the Internet or in a college library. The purpose of this book is to focus on some of the fundamentals that are key toward enabling the reader to build a foundation in understanding how to solve a basic power integrity problem without having to resort to modeling in a CAD tool—before that basic understanding ever takes place. Thus, the objective here was to focus on the tools and the methods of the problem—rather than on the *tuning* of the solution to the problem—which is where many good CAD programs excel.

Thus, in that spirit, I set about crafting this book with a few basic goals in mind; first, introduce the concepts of power integrity by utilizing basic analytical tools. Second, structure each chapter so that the complexity increases (for the most part) as one progresses further into the text. Third, emphasize the ability to set up problems—without the use of advanced software programs—enabling the reader to grasp the concepts first before embarking on a complex modeling exercise. Finally, fourth, introduce power integrity from a *systems* perspective rather than focusing on just the network analysis—which appears to be where many texts on power integrity tend to start, and sometimes stop, their learning paths. I hope that the reader will find that I satisfactorily accomplished these goals and that the information within the text is useful.

ACKNOWLEDGMENTS

There were many people who have guided me over the years, and I wish I could thank them all explicitly here. To those who are not mentioned below, my gratitude is given nonetheless.

To my editors at Wiley, for their patience through my wife's illness. To my two long time mentors and friends David H. Hartke and Dr. James L. Knighten for their tremendous faith and guidance throughout the years—and thanks again Jim for the Foreword and edit suggestions as well! To Dr. Keith Muller—words would not suffice here to express my thanks. To the late, great, Dr. Clayton R. Paul for his faith in me and support. To my friend Joseph S. Riel for his brilliance and insights in so many ways. To Dr. Jack Shemer for his incredible business teachings and leadership. To Dr. David Hochanson, for those long deep talks. To Dr. James L. Drewniak for his help and insights thoughout the years—and, of course, to the UMR team. To Dr. Kevin Quest, my friend and advisor. To Dr. Henry Koertzen for his depth of knowledge in power technologies and edit suggestions. To my father and sister for our talks. To my friends Bob Fite and Ed Stanford at Intel. To my team at Intel—you know who you are. To my son for his patience—especially all those nights we had to miss. And finally, to my wife, to whom this book is dedicated, and who has graciously stood by throughout while remaining so amazingly strong under some very tough circumstances. I thank you all.

ACRONYMS

PI power integrity
PDN power distribution network
VR voltage regulator
TD time-domain
FD frequency-domain
AVP adaptive voltage positioning
VID voltage identification
DAC digital to analog converter
RSS root sum of squares
PCB printed circuit board
MOSFET metal oxide semiconductor field effect transistor
SSN simultaneous switching noise
FCC federal communication commission
OATS open area test site
RF radio frequency
DUT device under test
SA spectrum analyzer
FFT fast fourier transform
SoC system on a chip
MD mask designer
PMIC power management integrated circuit

1

INTRODUCTION TO POWER INTEGRITY

This book examines the design concepts of power delivery to modern microprocessors and other related high-speed silicon devices. Today this field is termed power integrity. This chapter provides the background information on what has driven the need for platform power integrity analysis in this relatively new field. The platform is essentially the computer board with its multiple silicon devices, in addition to the power sources, or *converters*, that power them. The subject of power conversion will be examined, in particular as it applies to areas relevant to power integrity engineering. For computer systems the power conversion is mainly in the DC–DC conversion area. The chapters that follow will discuss areas relevant to power integrity analysis—circuit and field theory, modeling, the power delivery network (PDN) and boundary analysis, and other system considerations—and end with an examination of system noise, loadline, and measurement techniques. The last chapter will introduce silicon power integrity, along with some advanced interrelated topics, because of the increasing interest now being given to silicon-level power and the problems associated with on-silicon and on-package power delivery.

In the present chapter, power integrity is defined in terms of the paths that make up the PDN. The paths and all their components comprise the PI (power integrity) domain. A historical review of the voltage and current changes over time (using the microprocessor as an example) is provided to show how silicon has been one of the driving forces behind the need for such fundamental power integrity analyses today. The concept of *first principles* is discussed, because utilizing known equations and

Fundamentals of Power Integrity for Computer Platforms and Systems, First Edition. J. Ted DiBene II.

circuit analysis helps one gain insight into complex problems prior to embarking on sophisticated modeling with numerical tools. A discussion of the limitations and boundaries in power distribution analysis follows covering the circuit limitations (noise sensitivity and silicon process technology) of many advanced devices today that can compromise the accuracy of results.

1.1 DEFINITION FOR POWER INTEGRITY

Power integrity as a field of study includes *power conversion, power distribution analysis, circuit analysis*, and often the *package/board/silicon system analysis*. But PI is not limited to these subjects. The PI engineer should also be versed in thermal and mechanical basics because some problems needing to be solved may include these and components of other disciplines that impact the system under study. A simple, but somewhat limiting definition is:

Power Integrity *The study of the efficacy of the power delivered from the source to the load within an electronic system.*

Today, power integrity engineers versed in other disciplines may need to consider in their analyses the *system's source*, *load*, and *path*. In the past power integrity engineers often excluded the source and load parts of a system. This is understandable because many power integrity problems focused only on the power distribution path. However, today, having knowledge of both the silicon load *and* the power source allows PI engineers to comprehend fully the complexities of the problems that they face. Conversely, many power conversion engineers are required to cross over into the power integrity domain in order to solve their domain's problems. It is therefore reasonable for engineers from both disciplines to move regularly into each other's domains in order to solve their problems satisfactorily.

As Figure 1.1 shows, the primary power source is the power converter. This is a type of DC–DC converter at the motherboard of the server, forming an inter computer platform. The power source to this converter is typically neglected in an analysis. The power converter includes a certain amount of *decoupling* for filtering and charge storage. In the middle of the figure is the PDN, or *power distribution*

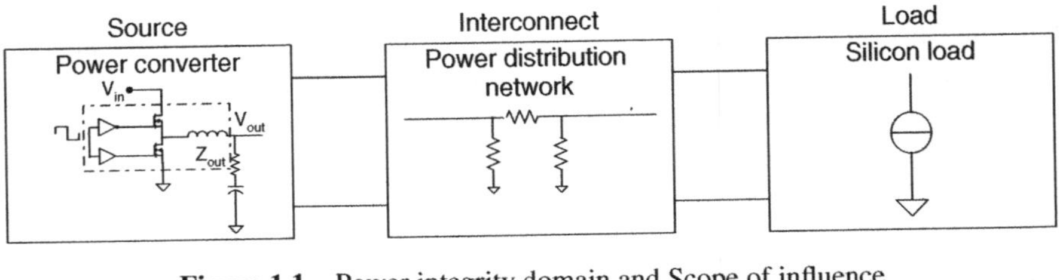

Figure 1.1 Power integrity domain and Scope of influence

network. The PDN typically comprises passive elements from the printed circuit board all the way to the level of the silicon. All of the decoupling and interconnections are included, from the output of the voltage regulator to the load. The printed circuit board and/or design package was where the PI engineer focused in the past. With the more complex recent systems, the PI engineer must often optimize the performance of the *silicon*-level passive and active components, so the circuit load must be considered as well. Note that Figure 1.1 gives a schematic representation of circuit load behavior. This is because modeling the actual behavior of load transistors under all possible conditions is virtually impossible. Nonetheless, knowledge of load behavior is required for PI engineers to do their jobs, and PI engineers must work closely with silicon development teams to gather the data necessary to perform their analyses.

For the PI engineer, at the start of a study, there are many complementary components to consider. Often these are thermal and even mechanical issues that contribute to delivery problems. It is then up to the PI engineer to rework the filter structures in the PDN and, together with the board and silicon teams, to ensure that the power delivery path is performing efficiently. Such analysis requires knowledge of the board's layout, its components, the power source, the load characteristics, the design package, noise coupling to other planes, EMI (electromagnetic interference) issues, and other items that may contribute (potentially) to the results of PI modeling. The assumptions that go into the analysis are a critical part of a PI engineer's responsibilities and the next chapters will explore in detail these assumptions.

1.2 HISTORICAL PERSPECTIVE ON POWER INTEGRITY DRIVERS

The idea of analyzing the power distribution path is not novel. Engineers have been working with the concept of measuring voltages and currents on power lines since the 1920s [1]. However, the need for advanced power integrity techniques in the computer was not realized until recently. The transition from virtually *no* power integrity analysis needed to its being *required* on virtually every platform developed today has been more dramatic than many technologists could have realized. Though noise, EMI, and signal fidelity have traditionally been areas of focus for the system designer, the need for advanced power integrity analysis, relative to the advent of the microprocessor, is still a recent event. Many conferences today are dedicating significant blocks of time to the multitude of papers written on the subject in just the past few years. The main factors behind this trend are a culmination of system metrics: the need for more stringent voltage and current requirements, the increase in voltage rail proliferation (internal and external to the silicon), a dramatic increase in platform and silicon signal densities, and device and platform cost pressures, to name just a few. As evidenced by the previous issues, many of these developments are clearly platform dependent. The issues though vary across each platform type. For example, voltage proliferation and current and voltage requirements may be the main issue on a server platform, whereas cost will typically dominate in a desktop, laptop, or tablet today. Nonetheless, the problems are very similar between them and the migration toward a deeper understanding of the state of the art is clearly needed today.

One very dramatic change that has occurred over the past two decades is in silicon power requirements. The relatively fast decline in source voltages and the sudden increase in currents delivered to silicon over this time has necessitated a stringent examination of the power delivery path—particularly for complex devices, such as microprocessors [3].[1] Figure 1.2 shows the voltage decline over time for microprocessor devices for the past 20 years. The changes in supply voltage have come about for a number of reasons. First, as complexity and transistor density has increased, so has the power required for the device. This has necessitated that the voltage be dropped to help reduce the overall power to the device. Second, as the silicon process geometries have shrunk, so has the requirement to reduce the voltages to the devices to prevent their damage. Thus manufacturing constraints and device physics have also driven the need to reduce the voltage to these devices as much as any factor. This has in turn driven the power converter industry to follow suit to supply suitable power to these complex devices [2]. The simple graph in Figure 1.2, however, may be a bit misleading. The voltages to these devices have *indeed* been reduced over the past decade. However, today, many high performance silicon devices and the *system on a chip* (SoC) devices are highly integrated, meaning they have other functional units in them besides the processing units. Thus this graph represents—for the silicon parts in more recent years—merely the voltage trends for the processing *elements*. For example, many processors have multiple processing *cores*, IO, memory, graphics, and power management units in them as well. The implication here is that most high-performance silicon chips now require *more* than one supply voltage to power these different units, and most of the supply rails for these units require *different* voltages.

The other important development to note is the *lower* core voltages in use today compared to just a mere 20 years ago. This change alone has led the rise toward power integrity analysis. Today, most core voltages are at or below one volt.

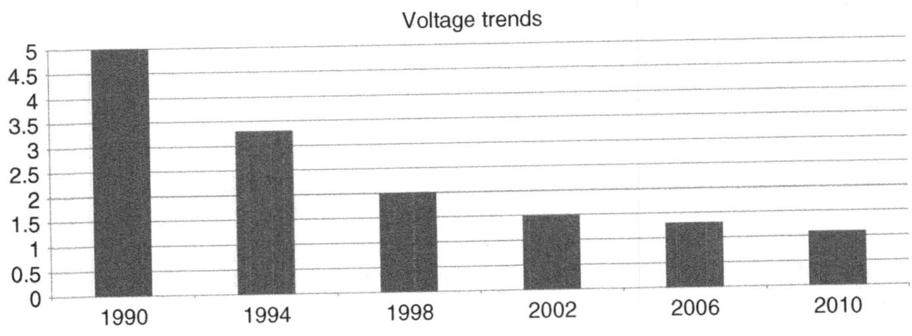

Figure 1.2 Microprocessor voltage trends over the past 20-year period

[1] The trend is not limited to microprocessors. Many other high-performance devices had also been operating at these voltage levels at this time.

This means that even the smallest noise voltage on the power bus may influence functional operation and cause data corruption. In 1990, the 5-V power rail to the processor (or the high-performance computing device) was typically specified at +/−5%, or +/−250 mv. Over time the tolerance for this power source has decreased to a voltage setpoint of only +/−1% today. Most complex devices require this tight regulation to ensure proper frequency and efficacy of operation. If a supply voltage were varied over the 500-mv range (+/−250 mv), for a 1V rail the voltage tolerance would be today +/−25%! Clearly, it is highly unlikely that the device would even function under such conditions. Thus, the noise margins (by necessity) have been required to shrink along with the voltage source to ensure the correct operation of the processor. This has made the job of the power integrity engineer that much more important.

An interesting change was also made in *current* over this same period; see Figure 1.3. The increase in current is graphed in a *log scale* to show the progression of growth. Notice that the *peak* currents[2] have increased by more than *two orders of magnitude* over just one decade.[3] So dynamic was this change that independent groups and even whole *companies* were created to help with the high current distribution problems that affected the motherboards and processing systems.[4] This change *alone* has necessitated the need for more expertise in power system and power integrity analysis.

Figure 1.4 gives a pictorial representation of this evolution. In 1990, the typical personal computer had noncolor screen (though color screens were coming into use at that time), a large ATX box that sat under the desk, and a large dot-matrix printer. Wireless communications in the office were either not available or

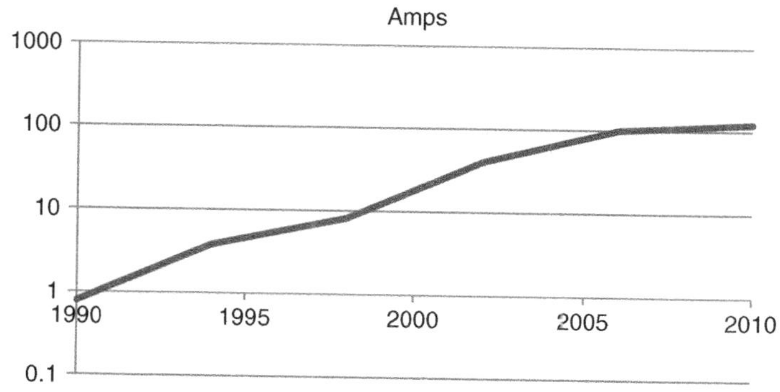

Figure 1.3 Peak current trends for high-performance silicon over the past 20-year period

[2]This trend does not represent the continuous, or steady-state current changes. These are typically lower than the peak currents.

[3]The current trend data are highly dependent on specific devices and are not representative of every device. The trends are also independent of manufacturer and considered generic.

[4]Many of the companies are focused on delivering CAD software for analysis purpose. A simple search on the internet on the subject of Power Integrity reveals a number of these.

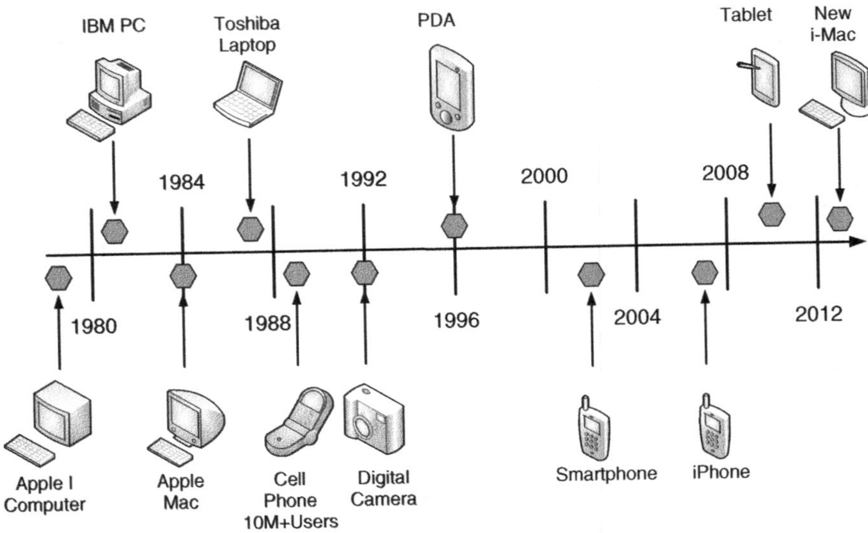

Figure 1.4 Evolution of modern computers

just starting. Today, many people use a tablet or smartphone to communicate in their personal and professional lives. The Internet has also revolutionized the way people use electronic systems.

What is *not* shown in the Figure 1.4 is the evolution of the *silicon* over the same period of time. *Moore's Law* has continued to hold true over this time window as well, as the number of transistors has doubled in computing hardware every 18 months. No doubt, as transistor density increases, so will the need to power these transistors with high-quality power.

The merge of communications and computers that has occurred over the past 20 years has changed the way people live their daily lives. This change has also motivated the development of power integrity as a separate field of study. While, in many ways, the field of power integrity is very new, it is expected to grow rigorously in the coming years to meet advances in silicon and platform power delivery. And, as technology progresses, so will the demand for well-trained power integrity engineers.

1.3 FIRST PRINCIPLES ANALYSIS

As computers have evolved, so has the software that runs on computer systems. For the developer or engineer, the modeling requirements have become ever more sophisticated with more options and methods to analyze complex problems. CAD tools that utilize SPICE or numerical solvers with advanced models to solve for electromagnetic fields have grown substantially over the past two decades. In fact, today, more engineers have become heavily reliant upon these tools to resolve PI problems.

Moreover, in today's fast-paced world of research and development, engineers are faced with solving complex circuit problems that require a strong background in other fields. That is to say, despite the fact that more sophisticated tools are now at their disposal, engineers cannot bypass this deeper learning and dive headlong into a simulation effort with the goal, inevitably, to deliver a solution that is satisfactory and meets the required deadline. The result is often error in the analytical process—or a less than optimal solution—and more lost time in the end to find a more adequate solution based on the first simulation. Those engineers coming out of school who lean too heavily upon the available tools and less upon the basics to solve difficult problems soon become frustrated.

However, the blame is not with the CAD companies and the ease in which their tools generate data. It is that professional engineers often lack preparation in analytical skills.

Not surprisingly, there is a good solution to this problem. In most cases, seemingly overly complex problems can be solved by understanding the *first principles* behind such problems before embarking on a detailed model and simulation effort. This does not mean that the final solution to the problem will be any simpler than the numerical modeling effort; it just means that the *estimates* and *boundaries* placed upon the problem (initially) will lead to a deeper understanding of the problem and thus a more efficient path to the solution.

1.3.1 Steps to Solve Power Distribution Problems

The *first principles analysis* is simply a way of using basic equations and physics (tools engineers have encounterd in their education) to describe a problem, and then using those tools to solve the problem. Usually the mathematical representations used to describe the problem are close *approximations* to the exact equations. For this reason, it is important to have the understanding that comes with the mathematics. Indeed, many complex problems can be solved quickly, using pen and paper, a simple spreadsheet, or a math program, and thus give the engineer quick insight into the boundaries of the solution. *First principles* will allow engineers to solve a related problem in order to gain insight into the solution of the problem of interest. While the results are typically an approximation to the actual solution, they are satisfactory for the initial analysis.

Invariably, the key for gaining a reasonable result to a complex problem is twofold: first, the assumptions must be within the bounds of physical behavior, and second, the limitations of the physical and mathematical models must be understood up front. The steps in solving a problem using the *first principles method* are straightforward and have been used by engineers in schools and in their professions for many years, though perhaps not known by this term. The steps outlined below are for solving a simple AC, or frequency-dependent, power integrity distribution problem. However, the same method could be used for any engineering problem.

1. Determine the result you expect to achieve, (Is this an impedance profile? Filter analysis? What is the output you expect to have?) Here, the objective is to capture the AC impedance across a given bandwidth.
2. Determine the relative positions of the components involved in the distribution and their dimensions relative to each other. Typically, this means sketching a plan and/or a cross-sectional view of the problem. For this problem, a cross-section of the various distribution networks may be drawn with placement of the components in the "plan view" of the networks.
3. Create a simple schematic of the problem. If the problem is too cumbersome or large, break it into subsections and analyze each subsection one at a time. Comfine them into larger sections and re-analyze. Here again, draw the network components (R, L, and C's).
4. Fill in the values for the schematic, starting with the known quantities (e.g., the capacitors).
5. Extract the interconnecting values from the basic equations and add these to the schematic to complete it. Here, one would use the tools of the later chapters of this book to extract these values, which are typically approximations to the exact values. Check to determine if each approximation is satisfactory for the desired estimation.
6. Determine, using simple hand calculations, a math program or, using SPICE, the solution, then examine the results.
7. Check the data against the simplifications. Typically, this means shorting or opening parts of the circuit to see if the results make sense. This does not always guarantee the assumptions or analysis is correct, but it is a good first check on the data.

This *first principles* method will work for most problems, such as time-domain analyses, with minimal modifications. It is also possible to combine it with numerical results obtained from data of other programs, published papers, and so forth. In any case, some simplifications will need to be made to link data to the simple model obtained by these from first-principles steps.

These seven steps are essentially a variant on the *scientific method.* So the process is clearly not new. These steps will always cut down the number of simulations and modeling iterations used because they provide the engineer with a strong foundation on where to start in the analysis process and give some insights that can improve on the early assumptions. Later in this book, some further tools that support this procedure will be discussed, along with some illustrative examples. It is, however, assumed that the reader has basic knowledge of the equations used in solving the type of power integrity problems illustrated previously. The rest of this section sets the *foundation* for the analyses used throughout this text.

It should be noted that it is not the intention here to replace detailed numerical modeling with the *first principles* method. The main purpose is to show how a strong analytical foundation can help the power integrity engineer solve problems more efficiently using these additional skill sets.

1.3.2 Limitations in the Analytical and Numerical Process

Numerical modeling is used extensively in power integrity engineering, and sophisticated CAD tools, such as SPICE (circuit simulator), Maxwell (field solver), and math tools, are an essential part of the engineers toolbox. However, recently engineers have been depending too much on these tools without understanding their limitations. Many 3D modeling tools are based on either the *finite element method* or the *finite difference/volume* method. Additionally, excellent advancements have been seen in the *PEEC* (partial equivalent element circuit) method in the past decade. All of these methods have their advantages and disadvantages—which are usually centered around accuracy, convergence, time, and cost. The main problem that they all share is in the assumptions that are made by the engineer doing the modeling. Moreover, this is less of a problem with the tool than it is with the user's ability to correctly utilize the tool. It is beyond the scope of this text to re-visit all the boundary conditions for these methodologies, the reader is encouraged to look over Chapter 3, where some of these methods are discussed. Here, some basic warnings are provided on both utilizing the *first principles method* and on reporting data using these tools.

Currently, there are limitations to what the power integrity engineer can analyze. Much of this is due to the modeling constraints—such as in the software and accuracy of the assumptions—that go into the analyses. As systems have become more sophisticated, so has the need to have better and more accurate sets of tools. When it comes to modeling a problem, the PI engineer is mostly interested in the physical representation of the circuit structure and thus the issue becomes how to represent accurately the complex electromagnetic behavior with simple passive elements. For most problems, a simple circuit approximation to the actual is adequate. For example, extraction of parasitic resistance, conductance, inductance, and capacitance of a circuit may sometimes be done by the *first principles method*, at least at the beginning of the analysis. However, in more complex cases, advanced modeling tools are required to adequately extract the values. This is where the tools, and the assumptions that go into the tools, become important.

Prior to performing a detailed modeling analysis—which may take weeks or even months—it is good practice to assess the boundaries of the accuracy of the effort—and if it is *sufficient* for the given objective. Too often, engineers start with the software and then determine that the assumptions they were working are inaccurate. Figure 1.5 illustrates where (typically) the accuracy coverage of the problem is and at which point in the process flow the accuracy limits will be available. Usually, a *first principles analysis* (point A in figure 1.5) will yield a 50% accurate solution; that is, if the exact solution is 2×, the expected accuracy will be *at worst* between 1× and 3×. This should be the goal for the PI engineer when starting out on a problem, such as the analysis of a power distribution network, to ensure that the accuracy of the results is well understood. In addition, it is surprising how often even a *first principles analysis* is more than sufficient for many engineering challenges. Moreover, if a *first principles analysis* is performed, the engineer should check the accuracy of the equations used to make sure that they are within these

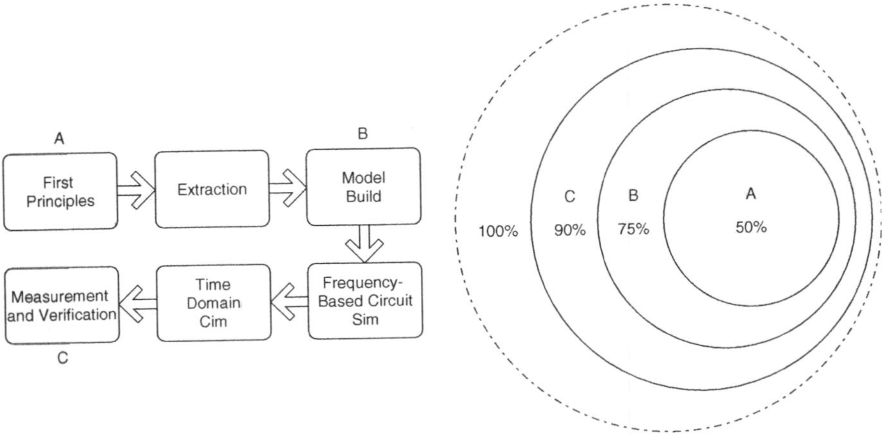

Figure 1.5 Coverage and accuracy chart for power integrity problem solution space

bounds. If the equation is, say, to *estimate* the inductance of an element in the path geometrically, this should be stated up front and the accumulation of the errors statistically should be noted at this point with the other components as well. As one progresses into the model building phase (B), the expected accuracy will typically improve—thus an error of 25% or better would be expected here. After this point in the analysis, it is usually good to assess whether or not more information is needed to make a decision on a particular project. If more accuracy is needed, the engineer will continue to the measurement and verification phase (C) where there is also a *tolerance*, but where an accuracy of at least 10% is sufficient. Next the data are validated to ensure that the load is accurate by using dynamic load tools and other instrumentation. Engineers may jump from one step to the next, sometimes bypassing a step along the way—provided that the data are sufficient from the previous step in the analysis.

Unfortunately, in practice, engineers too often report data based on inaccurate assumptions and analyses. For example, suppose that a power integrity engineer is asked to report on the droop deviation for a time-domain transient analysis in a PDN circuit. It is not unusual for engineering teams to have been part of such a review and to have reported the data in the following manner: the engineer completes the analysis, and a meeting is then called; then the engineer compiles the data in a format (e.g., a slide presentation) in order to share the data with the team. The final result is reported using a number based on SPICE modeling and the 3D extraction of the parasitics:

$$\Delta V = 7.849326 \text{ mV} \tag{1.1}$$

The first question one might ask is: What is wrong with this number? As will be seen shortly, this type of reporting is completely incorrect, but far too often this is how the data are presented. To begin with, based on the sphere in Figure 1.5, dimensionally, this much accuracy is not possible. Cleary one would expect the tolerance to be at best 25%. The engineer should have performed a simple dimensional analysis of the problem, to begin with. This way, the value has a *range* of accuracy—even if the simulation was perfect! The new number, simply based on the known tolerance of the elements (not even taking into account the other potential errors), would be

$$\Delta V \cong 8 \mp 2\,\text{mV} \tag{1.2}$$

Even if there were no errors in the simulation, the assumptions were perfect, the conditions were perfect, and the modeling was perfect, this would be the engineer's starting point. One might presume that the team would start to ask a number of questions if the data were reported in this manner. Unfortunately, it is common for the team to review the data and *not* question the accuracy, since it came from the tool, of course, and isn't the tool always correct! Many engineers do not even question the data until much later on in the process—often as late as when the lab measurement data comes back—at which time the data are predominately found to be very *different* from the simulations. At this point, gross errors in the results can cost significant money, time, and resources to recover. Invariably, this becomes problematic for the data owner, and those that may require the information. Moreover, reporting the data in this manner lacks certain clarifications, though it may look perfectly good to many on the first review. Missing in the analysis are some very fundamental facts that would lead a reviewer to question the result. With the present model, in a through review one would expect a number of questions to follow;

- How were the individual elements modeled? What were the assumptions in the model for the location of the ground reference for each parasitic?
- What was the tolerance for each parasitic? For example, if it was a plane inductance, what tolerance for the dielectric thickness was used, and what temperature used (for mechanical expansion)? What connections were assumed and what were the boundary conditions of the initial setup in the model?
- What was the mesh size and convergence criteria set to for the CAD program? How many iterations did it take? Did the engineer perform an accuracy assumption prior to running the simulation to determine what accuracy was expected just from the simulation?
- In the SPICE setup, what was the time-domain step size used? Did this person break down the PDN to a small enough lumped element so that step sizes did not matter?
- Was this analysis checked per an existing analysis so that a known baseline from either a measurement or a known simulation run would verify the results?

And the list could go on. These are just some of the questions that could come up in an actual review process. And *here* is where it becomes critical for the power integrity engineer—and for engineers in general—to *ensure* that the proper preparatory work is done on the problem. It may be worthwhile to put together a *checklist* to prevent errors in delivering such data. A few basic steps would be useful in the modeling and analytical process are:

Step 1. Start with an existing analysis that is similar enough (e.g., a known PDN) and look at the results of this droop event given the constraints. Most other engineers have gone down a similar path and have results that yield an excellent starting point—even so much that their initial models may be used.

Step 2. Review all of the parasitics—putting these in a table and looking at the tolerances of each separately is a good idea. The tolerances may be used for the SPICE analyses later on.

Step 3. Perform a *first principles* analysis on the PDN to make sure that, at least from a lumped element perspective, the data makes sense. Make sure that it is well understood that the accuracy of this first pass is a crude, though a reasonable, approximation to the final result. For example, if the droop result from this analysis is 100 mVpp, this would indicate that the 8 mVpp initial simulation may be suspicious.

Step 4. Set up the 3D modeling tool correctly, and once again, look for similar extraction results on already measured data if possible. Too often the setting up of the 3D modeling assumptions can result in gross errors that strongly influence the final results.

Step 5. Run the simulations based on the tolerances of every device and result in the system. This means that the engineer should run at least two boundary cases in addition to the nominal case. This will give the expected range of results.

These basic steps comprise a general approach to solving just about any problem. However, it is common to find that engineers can miss some basic steps in reporting out the data—and likely some errors in the assumptions as well. In this example a little more digging may reveal that the actual component values and tolerances, behind the result, that went into the engineers assumptions, may be quite different. This is not uncommon. In fact, after applying the basics to the earlier results—which were derived from an *actual* example—the new, re-run analysis resulted in the following data:

$$\Delta V \cong 42 \mp 13 \text{ mV} \tag{1.3}$$

Note here that the following missed steps occurred in the first reporting: the number of dimensions in the initial case, and actual result reported, were misleading and led to the wrong result. Moreover, the assumptions that went into the modeling effort were incorrect. This is why the process of generating data must be reviewed carefully to get a timely and accurate solution space.

As with all analyses, proper data representation and assumptions are not the only problems the engineer must face. For power integrity, in particular, an understanding of the environment affecting the power path is also important. The *boundary conditions* for the PI engineer must be considered prior to embarking on a complex and detailed analysis. The boundaries are typically focused around the *bandwidth* of the solution space. The bandwidth for solving power integrity problems is highly application and issue dependent, but the range for most problems is between DC and 1 GHz. This is a very wide range. Moreover, many noise-related PI problems have harmonics, and thus extending into this range and beyond must be considered—particularly at the silicon level. One could argue that within the silicon, power integrity frequencies extend well into the radio frequency regime. However, most problems fall into the noted bandwidth.

In summary, the critical action should be for the PI engineer to assess the proper boundaries of the problem prior to the simulation phase. This usually involves performing some basic computations, usually by hand or in a spreadsheet. What this does is set the level of complexity for the next phase of analysis and significantly speed up the process. That is, if the accuracy of the expected result is known *a priori*, then setting up the program to sample at a certain rate and through a given bandwidth will typically reduce the processing time. Understanding the boundaries of the problem will nearly always result in a more efficient and timely solution and will allow the engineer to be a more effective developer in the end. Chapter 4 provides some detailed examples that illustrate this process more thoroughly.

1.4 SCOPE OF THE TEXT

This chapter has provided a discussion of some of the key metrics in computer platforms that have driven power integrity and how the state of the art has evolved into an in-depth independent subject. The subject of boundary conditions was introduced and the limitations in power integrity associated with current modeling tools. However, the frequency bounds of power integrity will increase as we move forward into the next coming decades, since this is a dynamic environment.

Chapter 2 reviews power conversion devices used in computer system platforms. The discussion is necessarily brief, since there are many excellent texts on the subject for the student. The focus is therefore on power conversion as it relates to power integrity, rather than design. The review covers the basics of buck–level power conversion and how linear regulators operate. The discussion includes how to compute power losses and filter operation, how power switches work along with inductor and driver, some controller basics, as well as how current and voltage sensing operate. This introduction provides the baseline for later chapters where loadline and other subjects are discussed. Because the power conversion industry is also making advances in the area of platform power conversion some additional advanced subjects are briefly reviewed, namely coupled inductors along with multi-phase and tapped inductor topologies. The chapter ends with a discussion of platform-level power conversion, which will give the reader a basic understanding of the *source* block of the power integrity domain.

Chapter 3 reviews electromagnetic and circuit basics. Circuit extraction examples are introduced for various key geometries as a basis for the models used in later chapters. The simple first principle analyses will be valuable to the PI engineer. Some additional areas covered are power plane extraction, which is a type of problem the PI engineer may often encounter, and basic modeling methodologies commonly in use in software programs. To give any one of these subjects justice requires multiple texts, so the reader is referred to references at the end of the chapter for additional reading.

Chapter 4 introduces system PDN analysis. This is the middle domain region of the power path for the PI engineer. Within the power distribution network (PDN) are frequency-dependent impedance network distributions, and these are introduced with an emphasis on using first principles to analyze their structures. Some additional tools are presented, mainly concerning the printed circuit board and package structures, since a frequency domain analysis will change depending on which portion of the network is the focus. To supplement the analysis, connector interconnects will be discussed with an emphasis on socket type connectors, and in particular, contact resistance and how to treat this in the overall PDN. The chapter ends with a detailed discussion of decoupling fundamentals. This is an area where PI engineers lend the most expertise to the system.

Chapter 5 moves into the area of modeling the source and load. The emphasis is on the load, since much of the source was covered in Chapter 2. The focus is on the basic methods for modeling dynamic sources. Mastering this chapter's content will enable the engineer to generate a sufficient representation of the full path for time-dependent modeling, centered on the voltage bus droop analysis. This is likely the most important metric in power integrity analysis. Both closed loop and open loop methods are discussed and make use of modeling the buck regulator in SPICE. After the droop analysis distribution path losses are revisited for the entire link path in a discussion of bounding the overall power droop budget. Last, an *impedance boundary analysis* is presented that includes a frequency dependent representation to allow for a more complete mathematical view of the voltage droop function.

Up to this point in the book, the main elements of the power integrity domain have been covered. The focus then shifts to system considerations in Chapter 6. The important subject of the *power loadline* is introduced. The power loadline involves both the distribution network and the power converter. Next, signal noise and power noise on the power planes with the emphasis of coupling and how it is generated are covered. The next areas discussed are noise and signal representations in the PCB, at the package and silicon levels, with the focus on the different techniques that can reduce noise coupling in planes and improve power integrity distribution paths. Chapter 6 closes with a discussion of different PI measurement techniques.

Chapter 7 provides an introduction to silicon-and package-level power distribution and design futures. The discussion includes an analysis of the power delivery path inside of a silicon device and of the layout and routing issues related to the PDN. Relative frequency dependencies are discussed along with relative dimensions in devices such as advanced processors and SoC devices.

REFERENCES

1. Carroll, J. S., Peterson, T. F., Stray, R. Power measurements at high voltages and low power factors. *Trans. American Institute of Electrical Engineers*, 1924.
2. Stanford, E. Microprocessor voltage regulators and power supply trends and device requirements. *Proc. Int. Symp. Power Semiconductor Devices & IC's*, 2004.
3. Kwangok, J. A power-constrained MPU roadmap for the international Technology Roadmap for semiconductors (ITRS). IEEE SoC Design Conf. (ISOCC), 2009.
4. Singh, D., Rajgopal, S., Mehta, G., Patel, R. Technology Roadmap for Semiconductors. *IEEE Journals and Magazines*, 2001.
5. Tiwari, V et al. *Reducing Power in High-Performance Microprocessors*. Santa Clara, CA: Design Automation Conf; 1999.
6. Sullivan, D. M. *Electromagnetic Simulation Using the FDTD Method*. IEEE Press, 2000.
7. Reddy, J. N. *An Introduction to the Finite Element Method*, 2nd ed., McGraw-Hill, 1993.

2

INTRODUCTION TO PLATFORM POWER CONVERSION

Power integrity, as was discussed in Chapter 1, is a broad, and encompassing discipline. A key subject area is *power conversion;* this area is shown as the left-most block in Figure 1.1. Power conversion, even when focused on DC-to-DC conversion alone, is a complex and expansive subject, and to do it justice requires many years of study. This is why there are many excellent texts on the subject—some are recommended in the bibliography at the end of this chapter [1]. Thus, an in-depth coverage of power conversion—even limited to computer platforms—is well beyond the scope of this book.

Yet an introduction to this subject is important to those who study power integrity. Primarily, understanding the boundaries of the *source* is critical to the power integrity engineer's work. Without a basic understanding of the power converter behavior, it is impossible to follow the efficacy of the link in the frequency or time domain. Moreover, as this text moves to other subjects, the reader will be required to develop a basic knowledge base of the workings of the voltage regulator, and this need will become particularly evident when the subjects of *modeling* and *noise* enter the discussions. Additionally, it is necessary for the power integrity engineer to have a broader knowledge of the platform's inner workings, than just that of the distribution between the load and source. The PI engineer is often *required* to understand how the source functions in driving the PDN. Much of this learning comes from interactions with power converter designers, but this chapter will help with the basics of computer platform conversion that can lead to a good understanding of the *system* power integrity issues as a whole.

Fundamentals of Power Integrity for Computer Platforms and Systems, First Edition. J. Ted DiBene II.

Efficiency and power loss in a today's computer platforms is often one of the key metrics addressed by engineers, and not just by power converter designers. This is because energy efficiency is now a high priority issue for most electronic systems. For the PI engineer, having a good foundation as to what drives these losses in a complete power delivery solution, including the converter and PDN, is critical to dealing with some of the more difficult problems that can arise. In particular, it is the PI engineer who ensures that the PDN is *energy or power efficient* in the first place.

In summary, this chapter will introduce power conversion, with the main focus being on the basic understanding of the converter's functionality. DC–DC conversion will be covered[1] with an emphasis on *buck* converter operation. This is because the buck regulator is by far the most common converter in a computer platform. A few examples will be given in this chapter to illustrate computationally the effects of power conversion operation and power efficiency as well as power loss in a platform.

2.1 POWER DISTRIBUTION SYSTEM

There are essentially two main types of power distribution systems on computer platforms that are of interest here—*distributed* and *centralized* systems. A *centralized* power system is one where the power conversion is concentrated in a region and the power *distribution*, or each interconnect, fans out from this location. There can be only one to a few converters in a particular location and/or a single power device such as a PMIC (power management integrated circuit) that does the main conversion. Alternatively, a single voltage rail may feed the centralized block. This system is more common in smaller platforms such as in Tablets and Smartphones where the distribution network distances are smaller and the power rails' performance (speed, current amplitude, etc.) is typically less stringent than in higher performance platforms (e.g., *servers*). A *distributed* power system is where the power conversion is spread *throughout* the board and many of the converters are placed in proximity to the loads. Because the converters are located *away* from each other, they are fed by different power sources. These converters often operate independently; that is, with independent controllers. Both types are depicted in Figure 2.1.

2.1.1 Centralized and Distributed Distribution Systems

In both the *centralized* and the *distributed* systems, an AC-to-DC converter (often called a silver box or *brick*) is used to convert the voltages from an AC value to one or more DC values into the platform. Additional DC–DC converters then convert this main DC output voltage to other regulated voltages. In a centralized power system, the power is distributed from a centralized source to multiple loads or devices

[1] AC to DC conversion is usually handled outside the motherboard, and though is a critical subject, it will not be treated in this text except cursorily.

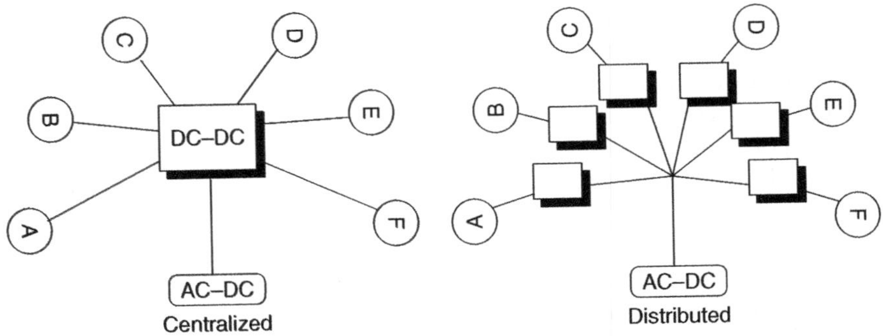

Figure 2.1 Examples of centralized and distributed power systems

(A through F in Figure 2.1). Often these are single silicon devices, but they can be multiple devices with one or more voltages sources delivered to them, depending on the load and device. In a distributed system, the DC–DC converters are *co-located* with the loads, the purpose of which is to provide higher quality power in proximity to all the device(s). It is not *always* necessary that the main power source be an AC-to-DC converter as well—some systems rely on a DC-routed bus voltage—such as a battery.

There is no right or wrong answer as to which system is best for one application or the other. A designer's choice typically comes down to a few key factors, which may include cost, size, and ease of integration, to name a few. The choice of power system is made very early in the development process. However, in today's high-performance electronics, usually one of the main drivers is *energy efficiency*. The power conversion and distribution network to the load plays a critical role in the efficiency of the power system. This is not only true for battery applications; it is now a dominant issue—even in high-performance server platforms—due to the rising costs of energy. Thus the engineer involved with the development of the power system must also have knowledge of the energy usage throughout the platform and must strive to make it as power efficient as possible. Normally, the power integrity engineer is involved in many of these decisions due to his responsibility for power to be delivered efficiently to these high-performance loads.

Many loss mechanisms in the path *beyond* the converter losses contribute to the energy efficiency of the platform. The power distribution network plays an important role in the losses of the system, and through power management of the devices in the platform—including the power converters—the *operation* of the load controls the power loss in this system. Indeed, power management is now a key technology driving platform power and platform efficiency. These subjects will be discussed in more detail in later chapters.

2.1.2 Static Losses in the System Power Path

In power loss across the platform, there is an *implied* interconnect between the AC–DC supply and the next stage VR—and similarly between the second-stage

VR and the silicon device(s), as shown in Figure 2.1. Because the power loss in the first stage (the AC–DC output to the DC–DC input) may be dependent on the voltage and currents and application,[2] the focus here will be on the *second stage*, which will be a DC–DC converter of some type. Often the load device (e.g., a processor or SoC) requires multiple different converters to power different segments of the device. The switching nature of the load—at least the portions that have the largest dynamic changes in current—may also require the power delivery and power integrity to be given more consideration in the design due to the more stringent system constraints.

Today, virtually every electronic system and/or device is required to be more efficient than that of the last generation. One basic system requirement is in determining how much power is required to ensure that the device is powered adequately under both *dynamic* and *static* conditions. Figure 2.2 shows the static portion of a typical interconnect schematic between the VR source (DC-to-DC converter) and the load. In order to determine the power loss of this interconnect, the engineer must first evaluate the efficiency of the converter from input to output—meaning compute the output voltages and currents under normal operating conditions. Second, the engineer must know the resistance of the interconnect path at the temperature of operation. Third, the engineer must know what maximum load current the load will draw. The efficiency of the converter is given by the following basic equation:

$$P_{Out} = P_{In}\,\eta \tag{2.1}$$

where η is defined as the efficiency of the converter, P_{OUT} is the static *output* power delivered to the load, and P_{IN} is the static *input* power delivered to the converter. One may compute the static power loss in the interconnect path by knowing the resistance, the temperature of each section, and the static current that the load is drawing. In general, it is common to sum up the resistances in each element in the interconnect independently once the temperature dependencies are known. This is described by

$$P_{Int} = I_{Int}^2 \sum_i^n R(T)_i \tag{2.2}$$

where $R(T)_i$ is the resistance of the element in the interconnect of interest and is dependent on the temperature T. I_{INT} is the current through the interconnect. Thus

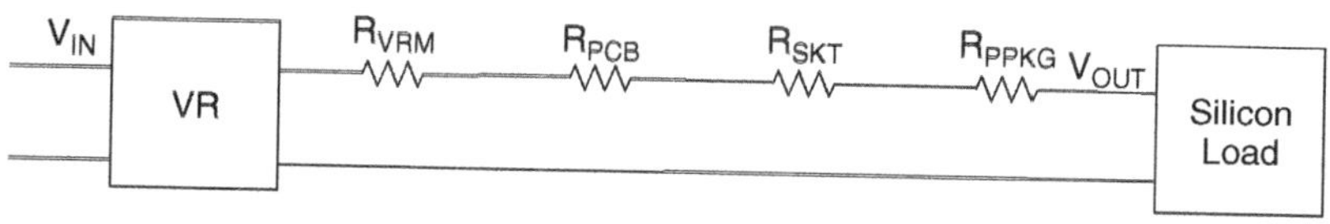

Figure 2.2 Simple DC model for distribution path

[2]There are many different front end designs—some require higher efficiencies and different voltages.

it is now just a simple computation to determine the overall power required for supplying the load current of the processor using equations (2.1) and (2.2). We use an example to illustrate this.

Example 2.1 A microprocessor requires a maximum of 80 W of power to be delivered to it at 1.5 V. The converter input voltage is 12 V into the DC–DC converter and is 86% efficient at maximum power and temperature. The DC resistance of the entire power distribution network (PDN) at room temperature (20°C) is 1.4 mΩ on *average*, and is made of copper. Estimate the total power loss in the system, the voltage at the output of the converter, and the converter input current and input power if the temperature of the interconnect is 80°C at maximum power. Assume that the temperature of each interconnect component is the same.

SOLUTION: The first goal is to find the voltage at the output of the converter. The resistance of copper changes approximately 0.43% per degree centigrade. Thus the *change* in resistance at 80°C will be

$$\Delta R_{80^\circ\mathrm{C}} = (1.4e-3)(80-20)(0.0043) = 360\ \mu\Omega \tag{2.3}$$

The voltage drop across the interconnect is determined by the current delivered to the load and the *total* resistance:

$$V_{INT} = I(R + \Delta R_{80^\circ\mathrm{C}}) = \left(\frac{80}{1.5}\right)(1.4e-3+0.36e-3) \cong 94\ \mathrm{mV} \tag{2.4}$$

Thus, to maintain a voltage of 1.5 V at the microprocessor (at DC only, when including ripple and other factors, this number will increase substantially) will require the output voltage to be approximately 1.59 V at the output of the regulator to account for the interconnect loss. The output power being delivered to the load is now the power delivered to the microprocessor and the interconnect:

$$P_{Out} = P_{uP} + P_{Int} = 80 + I^2 R_{80^\circ\mathrm{C}} = 80 + \left(\frac{80}{1.5}\right)^2 (1.76e-3) \cong 85\ \mathrm{W} \tag{2.5}$$

The input power may be determined from equation (2.1):

$$P_{In} = \frac{P_{Out}}{\eta} = \frac{85}{0.86} \cong 99\ \mathrm{W} \tag{2.6}$$

The input current is then simply the input power divided by the input voltage, which is 8.24 amperes. More often than not, the temperature changes in each resistive element in the path—depending on where and what it is—and these elements are typically separated out as in equation (2.2). This step is often the first step toward determining the *power* needs of the platform. The next steps involve examining the other system loss components and the AC or frequency-dependent losses in the system, which can be significant depending on the platform requirements.

In the example it is clear that both the *thermal* and *interconnect* constraints can affect the power consumption in a power path, and therefore can affect the *design* of the interconnect and the converter. It is important to make correct assumptions about how much power may be required depending on the system constraints and to understanding how the power is delivered to certain devices in terms of the overall power integrity of the platform. In this example, the efficiency of the converter was assumed *a priori*. Often this is not known up front, so an in-depth analysis of the converter, and its design, is required.

2.2 PLATFORM DC-TO-DC POWER CONVERSION

The objective of any power converter in a platform is to convert from a voltage that is not directly usable to a voltage that *is* usable. The conversion is required when the voltages transmitted from the main power grid, say, in a building, are not compatible with the loads; e.g. the points where the power is ultimately delivered. Unfortunately, there can be a huge discrepancy between the needs of the electronic load and the power from the substation. Fortunately, engineers have cleverly been able to develop ways to deliver very adequate power for the needs of these electronic loads throughout the modern age of electronics. For DC–DC converters, which take their input from an AC–DC converter (in most cases), this has typically been made up of topologies that make use of the higher voltage to drop down to the lower voltages distributed throughout the platform—as was showed earlier. Whether or not these circuits are of the *switching* variety, they all have one function in common: They transfer energy from one place to the other. Thus every DC–DC converter is basically an *energy transfer device*. The transfer function, from one voltage to another, is typically described by *the energy transfer relation* (ETR). This function describes how energy is *moved* from one place to another; it is simply the ratio of the transfer of energy from the input to the output of the system—which is similar to power transfer:

$$N_T = \frac{E_{Out}}{E_{In}} \tag{2.7}$$

where E_{OUT} is the output energy of the system and E_{IN} is the input energy in the system. The ratio is always less than one due to the losses in the system. The physical representation is shown in Figure 2.3.

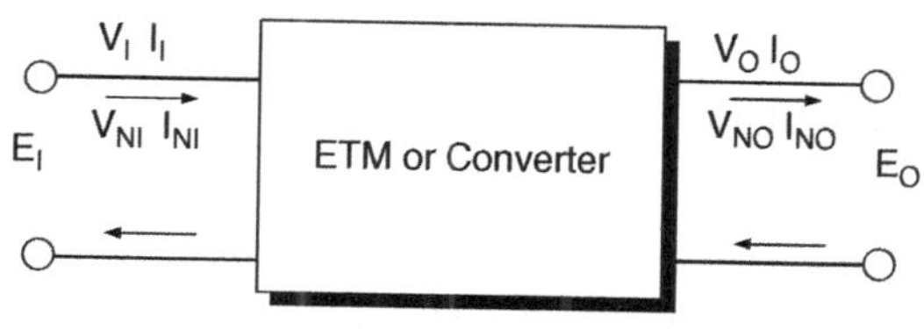

Figure 2.3 Simple energy transfer diagram

The ETR also is used to set the boundary conditions for the components, which helps with the functional transfer of energy from one place to the other. For example, if there are distinct bounds on the *noise* at both the input and output voltages of the system, using the ETR helps set the *parameters* of those components that comprise the actual converter. This is shown by the four noise parameters (V_{NI}, I_{NI}, V_{NO}, I_{NO}) at the input and output of the ETR. These parameters contribute to the function of the system and can affect the PDN design as well. They are particularly important for DC–DC systems where input and output noise are critical to the efficacy of the system.

In order to satisfy the relation in equation (2.7), the parameter N_T must incorporate the rate of transfer of energy from input to output, the boundaries on the AC signals at the input and output, and acceptable loss mechanisms in the system. Typically, these conditions are dynamic and must be satisfied over a range of values. The implications of this energy transfer relation for the buck regulator, in particular, are addressed by the VR designer and the power integrity engineer.

2.2.1 Popular Converter Types

In most advanced computer platforms today, the power conversion methods are as important as the quality of the power that is delivered to the load. For high-performance devices, the voltage regulator (VR) is located in proximity to the circuits and is designed to adjust to *point-of-load* voltages. The VR will continue to regulate within the specifications of the power delivery and load device under all normal operating conditions. VRs are mostly used in the *distributed* power system rather than the centralized structure. The motivation behind the use of this type of power architecture for such systems is mainly due to the evolution of electronics. The decrease in voltages of the load devices from generation to generation, along with the increase in current in high-performance silicon, has resulted in more complex power delivery constraints. As electronics have evolved, so have the power conversion and delivery path required to satisfy the changes. With the gradual decrease in the distance between the voltage regulator and the load, and the increase in the complexity of the design of the converter, the converter of choice today for delivering high-performance power has typically a type of *DC–DC buck topology* that converts higher input voltage to lower output voltage. There are variations on the buck topology as well as many other conversion methods on computer platforms, including linear, boost, buck-boost, charge pump converters, and isolated structures, but our focus will primarily be on the buck as it is by far the most common conversion topology on computer platforms today.

2.2.2 The Linear Regulator

A linear regulator is one of the simplest DC–DC converters in use today (Figure 2.4). It consists essentially of two components: a switch and an amplifier. The output voltage is regulated by a *controlled resistance* to the voltage drop across the main power switch when the switch is turned on. The switch operates

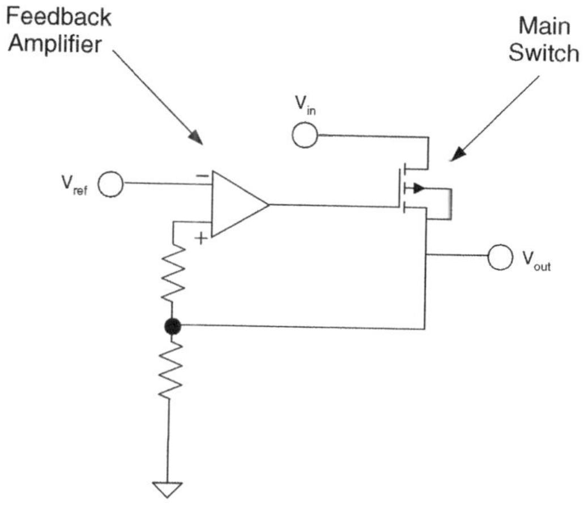

Figure 2.4 Simple linear regulator

almost exclusively in the *linear* region of the transistor, which is one reason why it is called a *linear regulator*. The other reason is that the voltage drop across the switch is mainly *linear* with the load—and thus with the power loss. The amplifier adjusts the voltage to the gate of the switch by sensing the output voltage through a feedback circuit. This voltage is compared against a reference voltage to the amplifier, and when the load changes along with the output voltage, V_{OUT}, the amplifier senses the change and adjusts the output voltage to maintain the regulated voltage setting.

The equations for determining the *efficiency* of the linear are straightforward. For the schematic in Figure 2.4, we get

$$P_{Out} \cong V_{Out}I_{Out} = V_{Out}I_{In} \tag{2.8}$$

The *approximate* term here is only the additional loss from the amplifier, which is often neglected in the computation because of its relatively small value compared with the main switch losses. Note that the input and output currents are the same since the switch directly connects the input voltage to the output through this *variable* resistor. However, this results in the power loss being *directly* related to the voltage drop across the switch as follows:

$$P_{Loss} = P_{In} - P_{Out} = (V_{In} - V_{Out})I_{In} \tag{2.9}$$

The equation for the efficiency is simply now the ratio of the output to the input power:

$$\eta = \frac{P_{Out}}{P_{In}} = \frac{V_{Out}I_{In}}{V_{In}I_{In}} = \frac{V_{Out}}{V_{In}} \tag{2.10}$$

To regulate the output voltage, the switch requires *headroom* in order to turn on properly. Thus the design of the amplifier and switch must always include some *minimum* voltage across the switch to regulate to the proper output voltage. This is the *difference* in the input and output voltages. Often this *voltage drop* is approximately 200 to 600 mV, depending on the linear regulator design, the silicon process, and the load to which it is delivering power.

The *advantages* of the linear regulator are that it is nearly always *monolithic* (all functions are located in one physical structure) and can be fabricated easily in a single silicon device and process and may be combined relatively easily with the digital circuitry. Moreover, because linear regulators have no switching components, the feedback loop is so fast that *noise* and voltage *drop* are reduced for most applications. This is why these devices are so well suited for *on-die* applications—where there is very little decoupling and speed and area is at a premium. The major *disadvantage* is that linear regulators are usually not very efficient powerwise and are often thermally limited for that reason. This is why these devices are typically used in applications where the voltage is relatively *low*, and *area*, not conversion efficiency, is often the premium metric of concern. The next example illustrates this aspect.

Example 2.2 An engineer is required to evaluate a linear regulator for powering a chipset on a workstation platform.[3] The requirement is that the regulator be at least 70% efficient, source up to10 amperes DC at maximum current, and operate with only board level cooling (meaning no use of a heat sink other than the PC board). The input voltage is 5 V and the output voltage must be regulated at two different voltage/current combinations: 0.7 V at 7 amperes and 1.5 V at 10 amperes. Determine what the efficiency, the power loss, and thermal constraints on the regulator are if it is to be designed to operate at a maximum of 150°C on this platform.

As part of the analysis, assume that the thermal resistance from junction to board (θ_{JB}), is 6.2°C/W, and the board ambient is 40°C at the maximum operating power loss for the linear.

SOLUTION: This analysis requires some very basic background in thermal analysis. For simple thermal constructions, a simple linear approximation (not to be confused with a linear regulator) is often sufficient to determine the thermal impacts of the system. In this case, the thermal resistance is given, which helps in the calculation. A few simple computations for this example should be sufficient to illustrate the point. The first goal, however, is to determine the efficiency of the linear for both operating points. Thus it will be necessary first to compute both the input and output power. From equation (2.9) the input power for each of the two conditions, is

$$P_{In\_1.5} = V_{In} I_{1.5} = 5 \cdot 10 = 50 \text{ W} \tag{2.11}$$

$$P_{In\_0.7} = V_{In} I_{0.7} = 5 \cdot 7 = 35 \text{ W} \tag{2.12}$$

[3] A workstation is typically a powerful desktop computer that often has a higher performing processor and other devices to run programs that are computationally intensive. The devices, for this reason, will invariably burn more power during operation, than say, a desktop or a laptop PC.

The output power will be

$$P_{Out\_1.5} = V_{Out}I_{1.5} = 1.5 \cdot 10 = 15 \text{ W} \tag{2.13}$$

$$P_{Out\_0.7} = V_{Out}I_{0.7} = 0.7 \cdot 7 = 4.9 \text{ W} \tag{2.14}$$

The efficiencies are simply the ratios of the output to input power for both modes:

$$\eta_{1.5} = \frac{15}{50} = 30\% \tag{2.15}$$

$$\eta_{0.7} = \frac{4.9}{35} = 14\% \tag{2.16}$$

Clearly, neither operating point is valid to meet the required efficiency, which was 70%. These kind of *checks* are usually warnings that the converter design is not well suited for the application. The power losses now for the different power constraints are

$$P_{Loss\_1.5} = 50 - 15 = 35 \text{ W} \tag{2.17}$$

$$P_{Loss\_0.7} = 35 - 4.9 = 30.1 \text{ W} \tag{2.18}$$

These losses are significant for *both* modes, and without the addition of a relatively expensive thermal solution, the engineer may fail to get approval to implement the regulator. The final check is to determine the temperature of the junction for the device. This will reveal whether the solution is practical from a normal thermal operating point perspective. The device's temperature increase from junction to board will be the product of the thermal resistance and the power loss. The computations for the two modes allow a determination to be made as to which is the worse case—and, for this example, it was determined to be the 1.5 V operating *mode*. The temperature at the *junction* of the device must be below the 150°C target. For this mode and application, the *change* in temperature, or ΔT_{rise}, will be

$$\Delta T_{Rise} = \theta_{JB} P_{Loss(\max)} = (6.2°\text{C/W})(35\text{W}) = 217°\text{C} \tag{2.19}$$

This result by far exceeds the upper limit set for the device, even without adding the board ambient temperature. Adding this in, the maximum junction temperature now is

$$T_J = 40 + 217 = 257°\text{C} \tag{2.20}$$

Thus this temperature is the internal operating temperature from the board ambient to the silicon, which is 107°C above the acceptable operating temperature. *Clearly,* a linear regulator for this application is not suitable. Often it is better to examine other methods, such as a switching regulator, which has a much higher efficiency. The buck regulator discussed in the next section is intended to be used for such higher performance applications.

2.2.3 The Buck Regulator

As was mentioned earlier, the *buck* switching converter is one of the most popular converters. It used to be called a *down* converter. The term "buck" refers to the way the switches are activated in the sequence, as described in this section: alternately, top and bottom; as the energy is transferred into the storage elements, the output voltage goes *down* based on the duty cycle ratio. The buck converter is popular because of its simplicity, cost, size, and efficiency as applied to many power problems. Table 2.1 shows how it compares to the linear regulator for a number of metrics.

The buck converter is also quite versatile and can be modified, or controlled, with relative ease. Another great feature is that today there is a seemingly endless list of suppliers who support the building blocks for this converter topology in the industry, making it relatively easy for the power engineer to find the components for their particular application. In fact, the buck converter has only a few key building blocks: bridges, drivers (sometimes built into the bridges), inductive and capacitive storage elements, and the controller. There are also some miscellaneous components for the controller feedback and sense circuits, but these are essentially considered part of the controller or filtering.

Figure 2.5 gives a schematic of a single-phase buck regulator. The shown components are the *upper* and *lower* switches, drivers, the LC filter, and the controller to regulate the output voltage. The bridge typically has two or more *NMOS* power transistors. A *PMOS* transistor is sometimes used for lower power applications and different voltage inputs for the top-side FET as well. The drivers supply the signals and power to the gates of bridges to the proper voltages and usually have a reasonably high drive current capability. The controller is typically a *PWM*, or *pulse-width modulation*-based design, but it may also be fully or partially digital in design. The main storage elements are the bulk capacitor and the series inductor. More discussion on these components will follow later in this chapter and in subsequent

TABLE 2.1 Some relative trade-offs for buck converter topology

Metric	Relative Comparison
Efficiency	Usally excellent
Size	Average, depending on application
Ripple	Good, depending on application (e.g. multi-phase and frequency of operation)
Response time	Good, particularly for multi-phase
Ease of use	Excellent, large base of suppliers for components
Cost	Excellent, again, due to large base of suppliers
System integration into platform	Excellent, very easy to implement with many possible designs available

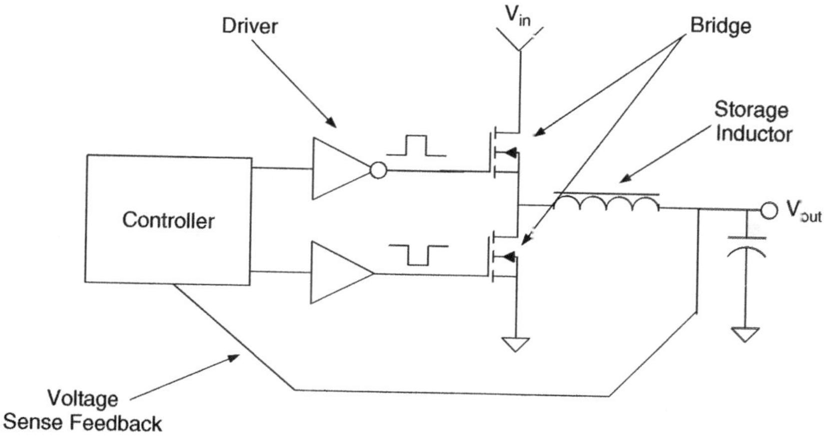

Figure 2.5 Diagram of simple single-phase buck regulator

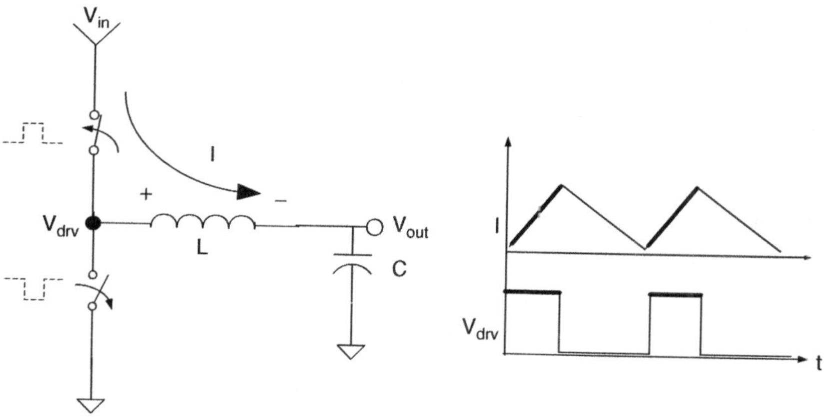

Figure 2.6 Buck converter operation in duty cycle mode

chapters. The operation of the buck is quite straightforward, and shown in Figures 2.6 and 2.7.

As shown in Figure 2.6, current *I* through the inductor increases when the upper switch is turned on. This coincides with the bridge node voltage, V_{drv}, which is the drive *mode* voltage between the switches. The voltage is *pulled* up to the input voltage V_{in} in this idealized diagram. When the upper switch is closed, and the lower switch is off (e.g., as shown simultaneously), current begins to flow into the inductor and increases linearly as shown in the figure. This time period is termed the *duty cycle* of the converter operation. The duty cycle occurs at the dark portions of the curves in Figure 2.6. When the switches are activated in *reverse*, as shown in

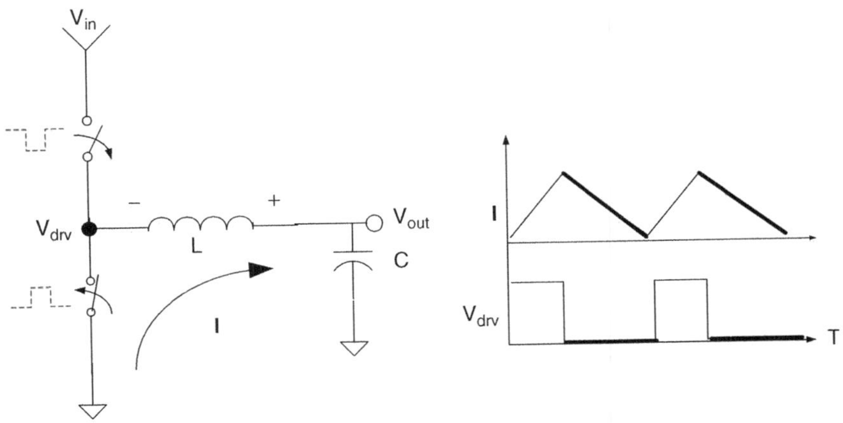

Figure 2.7 Operation of buck converter in commutating mode

Figure 2.7, the current continues to flow but the voltage *reverses* across the inductor (meaning the V_{drv} is now lower than the output voltage V_{Out}) to maintain the flow of current. This is termed as the *commutating cycle*.

In a buck converter, the duty cycle is directly related to the output voltage through the input voltage. In order to maintain a constant voltage (and stability), the volt-seconds across the inductor must be balanced. If they are not balanced, the converter will not operate correctly. In the figure, the period of operation occurs over the entire cycle T, with the output voltage determined as a function of the duty cycle. Thus, during the *duty cycle*, the voltage drop across the inductor will be

$$V_D = V_{In} - V_{Out} = L\frac{dI}{dt} \cong L\frac{\Delta I}{DT} \rightarrow (V_{In} - V_{Out})DT = L\Delta I \tag{2.21}$$

where V_D is the voltage across the inductor when the upper switch is *closed*. Conversely, during the *commutating* cycle, the volt-seconds must be the same across the inductor to ensure the correct operation. Thus the drop across the inductor will be

$$V_C = V_{Out} \cong L\frac{\Delta I}{(1-D)T} \rightarrow V_{Out}(1-D)T = L\Delta I \tag{2.22}$$

where V_C is the voltage across the inductor when the lower switch is on. Setting the equations equal to each other yields the equation relating the duty cycle to the input and output voltages:

$$(V_{In} - V_{Out})DT = V_{Out}(1-D)T \rightarrow D = \frac{V_{Out}}{V_{In}} \tag{2.23}$$

This equation states simply that the *duty cycle* of a buck regulator is directly proportional to the ratio of the output voltage to the input voltage. This is an important result, and is independent of whether the converter operates in a *fixed frequency* or *variable frequency* mode. It also reveals how *long* both switches need to be on in order to maintain the required output voltage—with respect to a fixed period T. As shown in Figures 2.6 and 2.7, the inductor current is directly related to these switch, *on-time*, durations. The converter may operate in this fashion in either *continuous* or *discontinuous* mode. When the current through the inductor is either at *zero* or above zero, the converter is operated in the *continuous* mode. If the current goes to zero, and there is a period of time where *no current* is flowing, it is considered *discontinuous*. Figure 2.8 shows the inductor current for both modes of operation.

As a side note, it is sometimes advantageous to turn *off* and *on* the switches at specific times in one or both of these modes in order to maintain the highest efficiency for the overall converter. This is called *soft-switching*, and it is a technique used by many designers—particularly with controllers that allow such programmability [1,2]. A converter is *soft-switched* when either the voltage or the current across one of the switches goes to zero (or close to it). Soft-switching is used when the efficiency of the switches are *non-ideal*, so the engineer may choose to increase the efficiency through this technique and/or others [3].

Designing for system efficiency, or more important, for the system to accommodate the *power loss* of the buck regulator was discussed in previous sections. However, determining efficiency for even a simple buck converter can be an arduous task. This is because there are many components that make up the power loss mechanisms in a buck regulator and power loss and efficiency vary as a function of load, voltage, thermals, frequency, and other operational features. Thus most designers have to perform the measurements in a lab to get results, and often the accuracy of the simulation and prediction process does not meet the required accuracy. Later in this chapter we will return to a discussion of the basic loss components of the buck regulator and examine the devices that make up the converter.

The next example shows how the buck regulator compares with the linear regulator from the previous example.

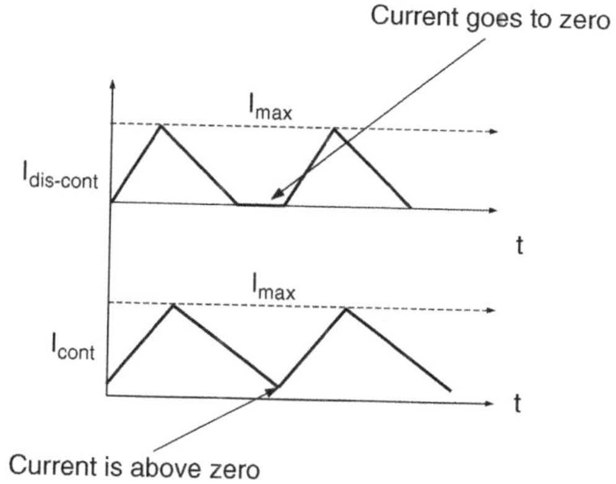

Figure 2.8 Discontinuous and continuous modes in buck converter inductor

Example 2.3 An engineer decides to evaluate a buck regulator for powering the same device as in Example 2.2. Assume that the same operating conditions as for the linear regulator. Compute the power loss for each operational state if the efficiency of the converter is 87% with 5 V_{In} and 1.5 V_{Out}, and 75% when it is 0.7 V_{Out}. Assume the same operating *loads* as before. Determine what the new thermal constraints are on the switches of the regulator if they are to be designed to operate at the same junction temperature as the linear regulator. In addition, compute the power loss in the main switches. Assume that there are two switches; also assume that the upper switch consumes the same power as the lower switch when the duty and commutating cycles are the same. That is, the power loss in each device is proportional to the time it is on. Assume that both switches consume together around 50% of the total loss in each operational state (the other losses are typically dominated by the inductor). Again, assume that the package thermal resistance is the same as the linear regulator with the same θ_{JB} (6.2°C/W) and same ambient board temperature of 40°C.

SOLUTION: The first objective is to compute the overall power loss for both operational states. This will give the overall power of the converter. The assumption here is that the main thermal constraint will be on the *switches*. In most converter designs, the inductor is also constrained, mainly because the designer often needs to optimize the design for size and cost. Thus, if the inductor is shrunk and the power increased, this component must be thermally managed as well. Fortunately, it is assumed here that only the switches must be dealt with.

The equation for the efficiency of a buck regulator is the same as it is for any regulator:

$$\eta = \frac{P_{Out}}{P_{In}} \tag{2.24}$$

The output power was computed in Example 2.2. Thus the input power for both modes is

$$P_{In\_1.5} = \frac{P_{Out\_1.5}}{\eta_{1.5}} = \frac{15}{0.87} = 17.24 \text{ W} \tag{2.25}$$

$$P_{In\_0.7} = \frac{P_{Out\_0.7}}{\eta_{0.7}} = \frac{4.9}{0.75} = 6.53 \text{ W} \tag{2.26}$$

And the losses for each are

$$P_{Loss\_1.5} = 17.24 - 15 = 2.24 \text{ W} \tag{2.27}$$

$$P_{Loss\_0.7} = 6.53 - 4.9 = 2.63 \text{ W} \tag{2.28}$$

Note that the power loss in the lower voltage output (0.7 V) is now much closer to that in the 1.5 V output case! The power loss in each switch is proportional to the on-time, or in this case the duty cycle for the upper switch and the commutating cycle in the lower switch. The next step is to compute the percentage of power in each as a function of this ratio.

For the 1.5 V output case, the ratio will be

$$P_{Loss\_Sw\_upper} = (\%tot)(tot\ pwr)\left(\frac{V_{Out}}{V_{In}}\right) = (0.5)(2.24)\left(\frac{1.5}{5}\right) \cong 0.34\ \text{W} \tag{2.29}$$

and for the lower switch,

$$P_{Loss\_Sw\_lower} = (0.5)(2.24)\left(\frac{3.5}{5}\right) \cong 0.78\ \text{W} \tag{2.30}$$

For the 0.7 V case the values are

$$P_{Loss\_Sw\_upper\_0.7} \cong 0.11\ \text{W} \tag{2.31}$$

$$P_{Loss\_Sw\_lower} \cong 0.70\ \text{W} \tag{2.32}$$

For the thermal conditions, the temperature rise in the worst-case thermal mode for the package is

$$\Delta T_{Rise} = \theta_{JB} P_{Loss(\max)} = \left(6.2\frac{°\text{C}}{\text{W}}\right)(0.78\ \text{W}) \cong 4.9°\text{C} \tag{2.33}$$

with a maximum junction temperature now of

$$T_J = 40 + 4.9 \cong 45°\text{C} \tag{2.34}$$

This is a more acceptable temperature for the devices on the board. It is true that though the power is significantly higher, the overall *area* for the power solution would be much smaller if the only component for power regulation were the linear regulator package. Thus, having multiple components allows the heat to *spread* out more easily with the buck regulator design. However, clearly the power consumption is reduced significantly in moving to a switching-based converter. The lower temperatures could allow the designer to use a less expensive package and let the switches operate at a higher temperature if needed. Practical considerations would nevertheless warrant a more detailed analysis of the size of these packages, which could limit the designer in other ways. Still, this simple example does illustrate an important point: sometimes the simplest solution is not always the best. The next section introduces some basics into the inductor and filter characteristics of the buck.

2.2.4 LC Filter Operation

The filter for the regulator is the key storage element for the buck converter. Though the input capacitance is also important, the inductor and capacitor is critical to the proper operation of the buck regulator. The LC filter stores energy and passes the low-frequency DC components of the *square-wave* signal generated at the V_{drv} node, as seen in the previous section. Essentially, the LC filter is a *bandpass* filter; that is, it is intended to pass only those signals within the frequency band of the LC filter. Because of this, practical LC filter designs cannot perfectly filter all of the ripple and noise generated from the square-wave power signal. Thus, it is expected that *some* of the frequency components at the switching node will pass to the output. This is typically the AC *ripple* through the inductor and is an important consideration in the design of any converter design. The equation for the resonance of the filter is simply

$$f_R = \frac{1}{2\pi\sqrt{LC}} \tag{2.35}$$

This is the resonant frequency of the filter *without* the parasitics added for each of the two elements. However, *all* filters have non-ideal components as shown in Figure 2.9. For this reason, it is important to consider the effects on the filter with the addition of the parasitics. Figure 2.10 shows the impedance of a filter with and without the non-ideal components added.

As expected, the *ideal* filter (solid line) impedance rolls off into the high-frequency band and the impedance drops 10 dB/decade after this resonance peak. This implies that the filter is no longer effective in passing these frequencies and will attenuate any voltages that are in this range. However, for the non-ideal case, there is an additional *null* in the resonance where the filter impedance no longer decreases and the impedance begins to increase again; thus, high-frequency noise may eventually pass through if care is not taken. This is why noise must be kept from being generated and coupled into the design of a voltage regulator, that is, the noise can affect the distribution path (PDN) and the silicon to which the power is supplied power. The inductive element—mostly from the addition

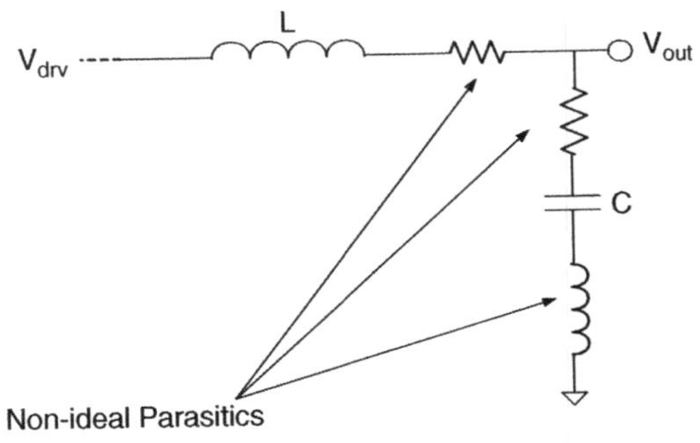

Figure 2.9 Typical parasitics for L-C filter

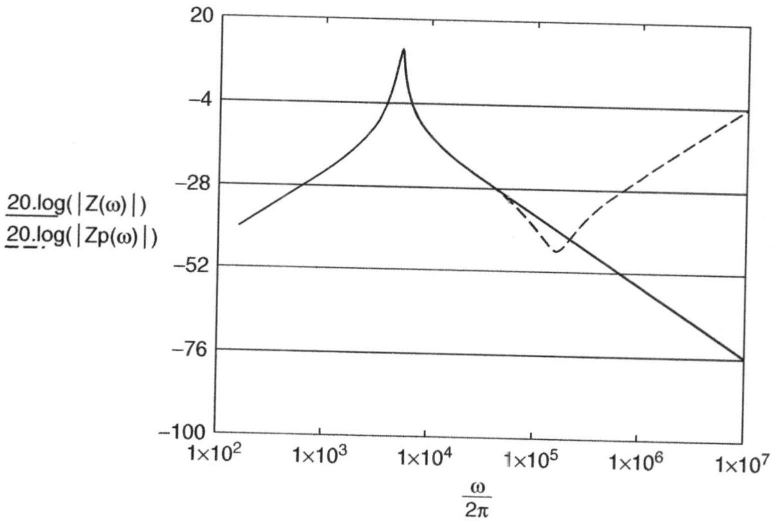

Figure 2.10 Impedance profile of output filter in decibels: With (Zp) parasitics and without (Z) parasitics

of the parasitics—can further affect the operation of the converter due to the voltage ripple of the converter. The ripple of the converter may be significant if the impedance is not sufficiently low enough within the band of the operating switching frequency. Because this is mostly a *reactive* power, the expectation is that this should not interfere with the overall efficiency of the filter. However, a large ripple can produce many negative effects at the system—one undermining its efficiency, the other its guardband. Thus, it is preferable to ensure that at the frequency of operation—and its harmonics—the ripple is kept in check. Guardband and noise are discussed in more detail in Chapters 5 and 6.

It is appropriate here to examine the operation of the filter, with a *square-wave* input representing the switching of the bridges, before discussing the ripple effect (Chapter 5). The voltage at the output of the filter is related to the voltage at the input via a *voltage divider*. If the eventual solution of the *transfer* function is to be in the time-domain, it is simpler to use a transform to solve for the output voltage first. Thus, the simple equation for this divider relation is

$$V_{Out}(s) = \frac{Z_C(s)}{Z_C(s) + Z_L(s)} V_{In}(s) = \frac{Z_C(s)}{Z_C(s) + Z_L(s)} V_{drv}(s) \tag{2.36}$$

where the use of a *Laplace* transformation is applied to represent the frequency dependence of the solution. The input voltage, in this case V_{drv}, is the *drive* voltage at the input inductor node where the bridges connect. The output voltage may be determined via the transfer function—assuming steady-state operation—to see how the filter affects this input signal. Thus, if it is assumed, for example that the parasitic R for the inductor and the parasitic L for the capacitor are negligible, the following equation is arrived at:

$$V_{Out}(s) = \frac{(R_C Cs + 1)}{s(LCs^2 + R_C Cs + 1)} \frac{(1 - e^{-sT})}{(1 + e^{-sT})} \tag{2.37}$$

where the Laplace transform for a square wave (*heaviside* function) of the drive voltage is used. A simple simulation will show the ripple due to this drive function (see Problem 2.7). Typically, to minimize the ripple, another stage of capacitance is added that has a different *series resonance frequency* (SRF) compared to the main storage capacitor. As shown in Figure 2.10, this is due to the additional parasitic resistance in the capacitor. Adding a capacitor with a *lower effective series resistance* or *ESR* often lowers the overall loss and ripple across the capacitor. The additional capacitance may also help *shift* the frequency response of the filter enough to mitigate the ripple further and reduce losses at the same time. It should be noted that if the inductor and capacitor are both ideal, no power will be lost in the filter because the amount of power that goes into the filter will be the same as that delivered. However, as was noted above, the parasitic components in both the inductor and capacitor contribute to an actual power loss that is very evident in an actual buck regulator. A discussion on capacitor analysis is given in Chapter 4, and inductor circuits will be discussed later in this chapter.

2.2.5 Power Switch Basics

Earlier in the discussion of the buck regulator, two main switching devices were used to create the drive voltage at the node of the bridge connection to the LC filter. These switches are usually what are called *power MOSFET devices*. The basic function of power MOSFET is to act as a switch for the transfer of energy from one part of the circuit—the input voltage in this case—to the next part—the output voltage. In the case of the *synchronous* buck regulator, this is the transfer from a *higher* voltage source to a *lower* one. Typically, the power integrity engineer is not required to understand the physics and detailed operation of the power FET (field effect transistor) in a switching regulator. However, in practical buck regulator designs, the power switch losses typically range anywhere from 30% to 60% in the regulator when supplying power to the load, and this is clearly a significant energy transfer loss, or impact to efficiency, in the design. For a power integrity engineer, having broad knowledge of a computer platform often includes an understanding of where the losses in the power delivery path come from, and thus of the operation and power loss components of the power FET.

There are many types of power FETs in the industry today [5]. As system constraints have impacted silicon devices and other aspects of the platform, so too have the designs and specifications been impacted on the power MOSFET. There are full texts written on the power MOSFET and there is plethora of information available to the reader on this subject. For this short discussion the focus will be on structures similar to the vertically diffused or VDMOS power FETs.

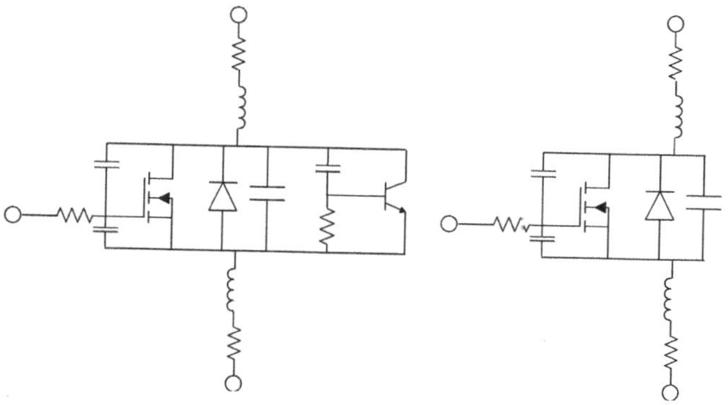

Figure 2.11 Two power N-MOSFET models—Complex and simple

Power switches today have fairly complex structures, and most electrical models are quite sophisticated. However, the switch model can be simplified enough to allow for reasonable power computations within a power converter analysis. Figure 2.11 shows two models, that are used for analyzing power dissipation in a typical power MOSFET device.

On the left in the figure is the more complex model (but not greatly so). For most power loss computations, the model on the right may be used to compute the AC and DC losses without loss in generality, and mainly for the capacitive *switching losses* in the FET and the static *on-resistance* losses, respectively. Additionally, while the upper and lower FETs physics are similar, their loss characteristics are somewhat different. A simplified equation for the losses of the device in Figure 2.11 is given below for both the upper and lower switches in a bridge configuration:

$$P_U = Static + Dynamic = Resistive + f \cdot (Turn\ on + Turn\ off) \tag{2.38}$$

$$P_U = I_{RMS}^2 \cdot R_{DS\_on} + f \cdot \left(\frac{Q_{SW}}{I_G} \cdot V_{In} \cdot I_D + Q_{RR} \cdot V_{DS} + Q_G \cdot V_G + \frac{Q_{OSS}}{2} \cdot V_{In} \right) \tag{2.39}$$

where P_U is the upper MOSFET. For the lower device, the equation for power loss is

$$P_L = static + f \cdot (Gateloss + Turn\ off + Body\ diode) \tag{2.40}$$

$$P_L = I_{RMS}^2 \cdot R_{DS\_on} + f \cdot \left(Q_G \cdot V_G + \frac{Q_{OSS}}{2} \cdot V_{In} + 2 \cdot V_{BD} \cdot I_{Out} \cdot t_{off} \right) \tag{2.41}$$

The parameters and definitions for equations (2.39) and (2.41) are given in the Table 2.2.

There are a number of MOSFET technologies available to the VR designer today. Most of the devices used in buck regulators are *N-channel* type rather than

TABLE 2.2 Parameter definitions for power mosfet models

Parameter	Definition
I_{RMS}	RMS current in resistive portion of MOSFET
R_{DS_ON}	On-resistance of MOSFET when turned fully on
f	Frequency of switching regulator
Q_{SW}	Gate-switching charge of MOSFET
I_G	Gate current
V_{IN}	Input voltage to device
I_D	Drain current of device
Q_{RR}	Reverse recovery time charge on FET
V_{DS}	Drain to source voltage across FET
Q_G	Gate charge on FET
V_G	Gate voltage on FET
Q_{OSS}	Output charge, gate, and shorted source
V_{BD}	Voltage on body diode
t_{off}	Minimum turn-off time on FET

P-channel, due to the lower on-resistance per unit area of device, which usually translates to lower cost. Many power converters on computer platforms utilize 12 V as the input voltage to the regulator. Thus, the devices must also withstand voltage deviations well above this for short time-windows in order to operate reliably. A few different popular types of devices used for platform VRs today are shown in Table 2.3.

The differences between the FETs are many; for example, the typical power loss for a GaN power FET may be much lower than for the other types, which allows for a smaller package in most cases. For the PI engineer, the goal is to understand and bound these losses within the platform so that the PDN design can adequately support the overall power converter design and load.

Computing the losses in the power switches can be an arduous task depending on the accuracy required. Often a designer must resort to detailed simulations in SPICE and even thermal/mechanical modeling tools, where with advanced CAD programs the precision may still be lacking. For the power integrity engineer, usually a first pass estimation is sufficient to acquire the necessary data. Moreover, the

TABLE 2.3 Power FET types commonly used for platform voltage regulation

Power FET Type
Trench—DMOS or VDMOS
Lateral—similar to VDMOS in operation
Gallium nitride (GaN)

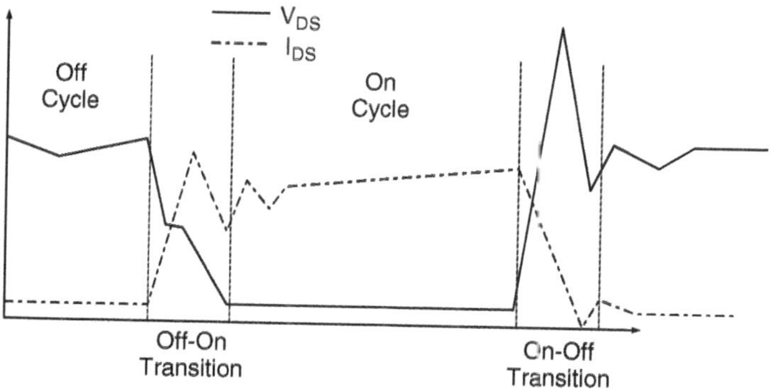

Figure 2.12 Example approximate piecewise linear current and voltage waveforms for power MOSFET

operation of the devices within the switching behavior of the converter can introduce *significant* losses that are not necessarily evident in the equations above. A typical waveform is shown in Figure 2.12. Note the *current spike* through the drain of the switch during the transition period, between the turning off and turning on of the device and the voltage spike on the opposite transition. This is a common occurrence, particularly between the transition period from the upper to the lower device turn-on and turn-off times. This usually necessitates that the control of the voltage regulator be some type of *soft-switching* control to minimize these *shoot-through* based losses.

The next example illustrates a simple computation for the power losses in the bridge of a buck regulator.

Example 2.4 An engineer wants to use a cheaper, older FET technology that is much more cost effective. The parameters for the new FET indicate that the performance is worse than the previous one and that the power loss and thermal resistance of the package will also be worse. Estimate the power loss during the commutating cycle for a MOSFET transistor given the equations in this section. The values for the MOSFET are given in Table 2.4 below. Assume that all of these parameters are correct for the temperature of operation. Also assume the same parameters as in Example 2.3. This means that the output currents and voltages are the same, and likewise the input voltage for the 1.5 V case. If the thermal resistance to the case (circuit board) is 25°C/W, (θ_{JC}), what is the actual temperature rise of the device internally during normal operation?

SOLUTION: Since this FET is operating in the *commutating* cycle, it is the power loss in the lower FET that is of interest. Equation (2.42) may be used to compute this power loss. The equation is repeated below, first to show what is missing:

$$P_L = I_{RMS}^2 \cdot R_{DS\_on} + f \cdot \left(Q_G \cdot V_G + \frac{Q_{OSS}}{2} \cdot V_{In} + 2 \cdot V_{BD} \cdot I_{Out} \cdot t_{off} \right) \quad (2.42)$$

TABLE 2.4 Data for power loss computation

Parameter	Value
I_1, I_2	9.5, 5.5 A
R_{DS_ON}	4.9 mΩ
F	500 KHz
I_G	400 μA
V_{In}	5 V
Q_G	25 nC
V_G	2.5 V
Q_{OSS}	2.5 nC
V_{BD}	1.25 V
t_{off}	125 ns

The RMS, or *root-mean-square*, current is one of the necessary parameters. If the current waveform is directly related to inductor current, then it can be assumed that the RMS current through the lower FET is related to the current by way of the inductor. The first step is to compute this RMS current. A simple drawing for this current waveform is appropriate and shown in Figure 2.13.

The equation for the RMS current is given as

$$I_{RMS} = \sqrt{\frac{1}{T}\int_0^T |I(t)|^2 dt} = \sqrt{\frac{t1}{3T}(I_1^2 + I_1 I_2 + I_2^2)} \tag{2.43}$$

This is the general relation for a continuous mode buck, as shown in Figure 2.13. The commutating time is found from the duty cycle and the frequency:

$$t_1 \cong \frac{1}{F}(D) = \frac{0.3}{500e^3} = 600 \text{ ns} \tag{2.44}$$

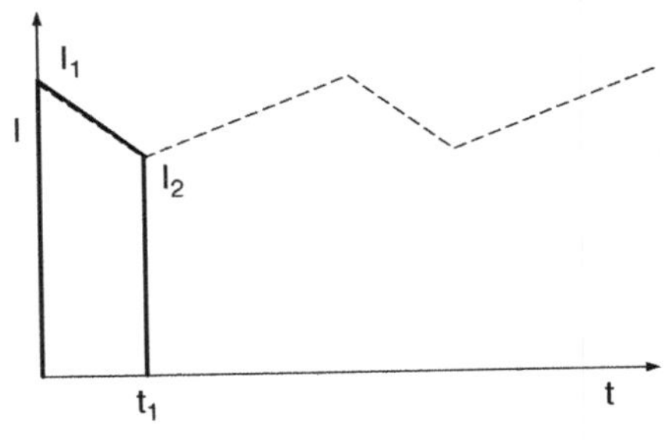

Figure 2.13 Current waveform for RMS computation

The solution is now

$$I_{RMS}^2 \cong \frac{t_1}{3T}(I_1^2 + I_1 I_2 + I_2^2) = \frac{(6e^{-7})(9.5^2 + 9.5 \cdot 5.5 + 5.5^2)}{3T} \cong 17.3 \quad (2.45)$$

The total power loss is obtained by just plugging the numbers into a spreadsheet. The computation may be checked against Problem 2.4 at the end of this chapter:

$$P_L \cong 1.7\ \text{W} \quad (2.46)$$

Again, it can be seen that the power is not significant for this FET as compared with the linear regulator. It is now a simple computation for the temperature rise:

$$\Delta T_R \cong \theta_{JC} \cdot P_L = (25)(1.7) \cong 42°\text{C} \quad (2.47)$$

If the board is at 40°C, this is well within the normal operating range of many power FETs at 82°C (typically these devices can operate safely up to 150°C). This is always good practice to check if the device has heated up significantly. A large temperature rise internally will also skew the power loss data, since the on-resistance and other parameters are usually highly temperature dependent.

As process silicon for logic and processing devices have evolved, so has the power MOSFET. The packaging technology, integration, and silicon processing have also advanced significantly, and have enabled smaller and more powerful designs for the power converter. The silicon MOSFET is an amazing device. It has evolved since its introduction in early 1980s, and it is expected to continue to improve and give power converter designers better performance, moving into the next decade and beyond.

2.2.6 The Controller

As was mentioned in the previous section, with regard to the power MOSFET, there are many texts that discuss the power converter controller in much detail. It is well beyond the scope of this text to discuss the controller for a power converter in the depth that is required to fully understand its operation. Thus, the discussion here will be cursory at best, and for more information, the reader is referred to the many references that are available. It is usually good though to start with a review of the fundamentals, and a good review is found in reference [5].

The controllers in use today for the buck converter range from the very simple single-phase *type I* feedback system to complex digital controllers for multi-phase voltage regulators. The type and complexity depend on the application. A little more than 20 years ago, many controllers were no more than simple integrated op-amps with PWM (pulse-width modulation) analog devices in them. Today, these devices often include sophisticated feedback systems that compute, either digitally or in an analog fashion, the voltage and current of each output of the VR and sense these

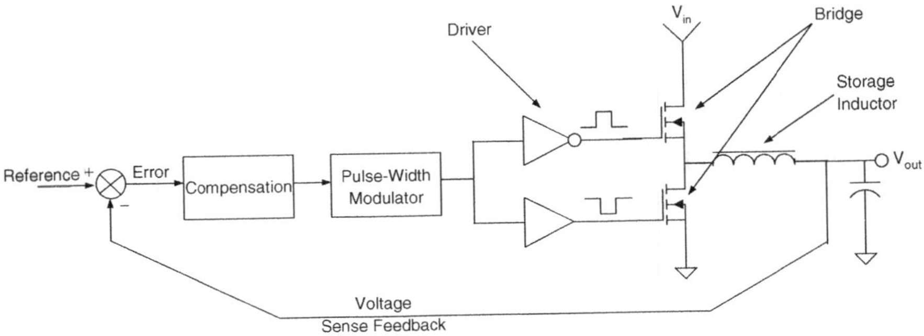

Figure 2.14 Simple buck regulator with controller and feedback

values at very fast speeds. Over the past 10 years, some suppliers have incorporated digital control into their portfolios that has allowed designers to develop voltage regulator systems with *nonlinear* responses to perturbations in the output voltage and current.

The basic schematic for a simple controller may be found in Figure 2.14. The controller is made up of components that allow the voltage to be regulated under all normal operating conditions. The system has a feedback signal at the output of the regulator, which is fed back to a junction node.

This signal is compared against a reference signal, which then generates the error signal for the compensator. The compensator adjusts the control signal such that the correct drive will be applied to the PWM. The PWM modulates this signal and sends it to the drivers, which amplify it sufficiently to control the bridges, and thus the power delivered into the filter and load.

The compensation network is normally an op-amp with feedback and passive elements that compensate for the changes in the output voltage. A type 3 compensator is shown in Figure 2.15. The compensation network and amplifier are not required to be analog in nature. As was discussed, there are many digital controller designs that emulate this function and incorporate it into complete digital designs.

Chapter 5 discusses how to model a simple controller for use in power integrity related analyses. The PWM generator typically comprises a triangular or sawtooth

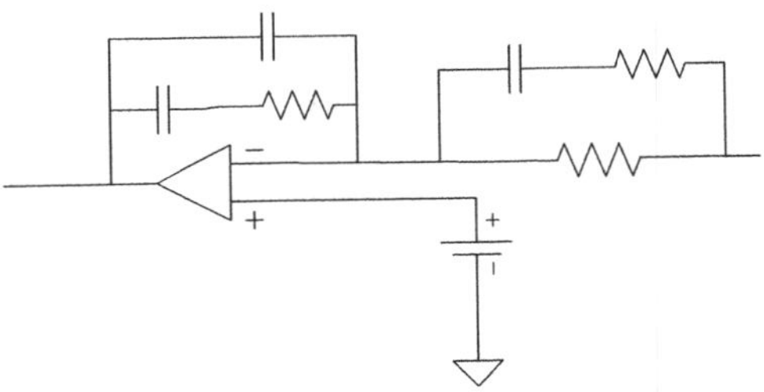

Figure 2.15 Simple type 3 compensation network

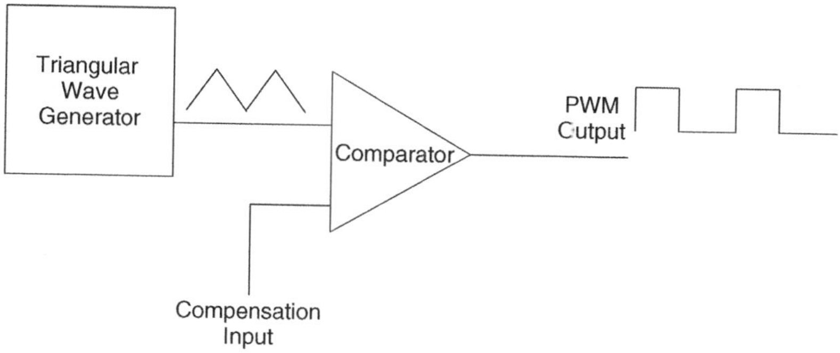

Figure 2.16 Simple PWM block diagram

generator and an amplifier/comparator that converts the triangular wave to a duty-cycle control signal that goes into the driver (see Figure 2.16).

The triangular wave generator can be adjusted by feedback from the compensation loop whereas the triangular wave *amplitude* is increased or decreased depending on the control signal into it. There are many different designs and methods to be used for closed loop systems. For the power integrity engineer, this introductory review, along with the modeling examples in Chapter 5, should be part of the toolbox required to begin analyzing basic power delivery systems.

2.2.7 Inductors

The inductor is the main storage element for the buck regulator, and it works to create a DC signal when connected with a capacitor, as discussed previously. The inductor acts as both a storage element and a filter. The energy stored in an inductor may be found by integrating the current times the voltage ($V \cdot I$), or instantaneous *power*, over time,

$$E_L = \int_{t1}^{t2} L\frac{di}{dt}Idt = \int_{i1}^{i2} L\,Idi = \frac{1}{2}L(i_2^2 - i_1^2) = \frac{1}{2}Li^2, \quad i_1 = 0 \qquad (2.48)$$

This expression is well known to engineers. If the inductor is quite large, all else being equal, then the energy that can be stored in the inductor will also be large. Conversely, a small inductance will limit the inductor's capacity to store energy.

This is the general case for the energy stored in an inductor. The inductor current resembles a triangular wave, and thus, it is a simple matter to solve for the energy in the inductor by integrating over the period (duty plus commutating cycle) on the buck regulator.

Earlier in Section 2.2.4, the LC filter for the buck regulator was discussed. As was noted, the value of the inductance and capacitance determines the magnitude of the AC signal (ripple) that passes through the filter when examining the impedance of that filter. For *low*-frequency switching regulators, larger values are required to attenuate this ripple. For *higher* frequencies, it is the opposite. The benefit of switching at higher frequencies is that—for a buck regulator—the sizes of the inductor and capacitor may shrink, which is often of great value to the VR designer. Chapters 5 and 6 go into the effects of ripple and noise in a system and how these artifacts can impact the development of the PDN.

Unfortunately, as the frequencies increase, so do the losses in the inductor and the capacitor. Capacitor losses will be discussed in more detail in Chapter 4. The *power* inductor experiences both AC and DC losses, whereas the capacitor—for the typical platform buck regulator—is usually only *AC* loss dominated. The DC losses for the inductor are fairly straightforward and are made up of (primarily) the metal wires that go through the inductor. The AC losses, however, are more complex and are dominated by the characteristics of the magnetic material that typically surrounds the wire (for most high-performance VR power inductors). An inductor construction along with a simple model is shown in Figure 2.17. Only the DC inductor resistance is shown in this model.

The physical construction may have two terminals on each side to accommodate higher current (as shown in the figure) or just one on each side. The DC component for the resistance typically has two parts—one external and one internal. The external part connect to the board while the internal part has one or more wires that egress into and out of the inductor material. The AC component comprises the losses due to the magnetic material. Additionally, if the frequency of operation is high enough, there will be an AC *loss* component due to the metal portion, which is due to the skin-effect. This subject is discussed some more in the next chapter.

The DC losses for an inductor may be described as

$$P_{L(DC)} = I_{DC}^2 R_{DC} \tag{2.49}$$

The resistances are typically given by the vendor and the losses are often computed with the aid of the information provided in the data sheet. The AC losses consist of *magnetic* and *nonmagnetic* losses. The AC magnetic losses are predominately *hysteretic* and *eddy current* based, depending on the inductor material. Hysteretic

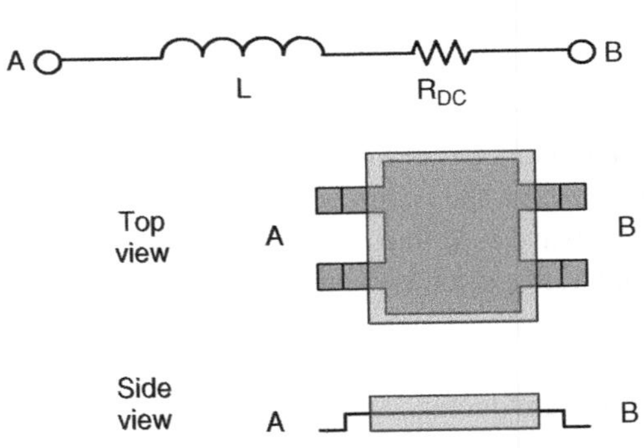

Figure 2.17 Examples of an inductor schematic and construction

losses may be estimated as

$$P_{L(Hys)} = f_S V_M 2 H_C \Delta B \tag{2.50}$$

where f_s is the switching frequency of the converter, V_m is the volume of the magnetic material, H_c is the *coercivity* of the magnetic material (this is the point in the B–H curve where the magnetic flux density B is zero), and ΔB is change in the magnetic flux density (magnetic flux density is described in Chapter 3). Hysteretic losses occur in magnetic structures when the magnetic domains expand and contract within the material and become *misaligned* temporarily due to the defects in the structure causing energy to be given off as heat. It is usually easier to account for these losses by examining the graphs from the data sheets and or by applying the data from a *Steinmetz* equation.

Eddy currents occur in magnetic structures where the current flows *orthogonal* to the magnetic field. Eddy current losses can be estimated using

$$P_{L(EC)} = \frac{l_m w_m t_m^3 B^2}{24 n_{lam}^2 \rho_m} \tag{2.51}$$

The parameters (l, w, and t) are the geometric variables for a structure—in this case it is assumed to be *rectangular* in form (e.g., having length, width, and thickness). The parameters n and ρ are the number of *laminations* and the *conductivity* of the magnetic material, respectively. At higher frequencies, for many *ferro*magnetic materials, both the hysteretic and eddy current losses can dominate the DC losses, so these need to be considered in the loss computations. However, for most ferrites, eddy current losses can be neglected.

As mentioned above, there are other AC losses that can contribute to the overall inductor losses. Two other types of losses, which are considered in magnetic structures, and even in *nonmagnetic* structures, are the *skin-effect* and *proximity* losses in the metals. The skin-effect is discussed in the next chapter. The *proximity effect* is similar to the current-crowding phenomenon except that it is *electrical*. A full discussion of these effects is beyond the scope of this book.

Often what is neglected in these simple computations (with respect to power loss) are the *harmonics* of the current waveforms through the inductor and how that affects the loss estimates. Typically (as was discussed in Chapter 1), these *first principle* calculations are not intended to replace detailed simulations or measurements. Nonetheless, their proper use is essential even if an estimate is the expected outcome.

When estimating the losses in a power inductor, it is most often easier to use the data supplied by the manufacturer of that inductor. Many suppliers generate data that allow a simplified computation of the losses in the inductor, particularly the AC losses. The supplier may even supply coefficients for the *Steinmetz parameters* of the magnetic structures. Steinmetz parameters are essentially an empirical representation of the complex nonlinear behavior in the magnetic structures and are found by *curve-fitting* data to a set of coefficients or parameters. The Steinmetz equation for a simple magnetic structure is

$$P_{L(SM)} = kf^{\alpha}B^{\beta} \tag{2.52}$$

where k, α, and β, are coefficients supplied by the vendor for computation of the losses at a given AC frequency. An example computation of an inductor power loss using the Steinmetz parameters is provided in Problem 2.6.

2.2.8 Coupled Inductors

Coupled inductors are becoming more and more common for power conversion applications. This is mainly because they help reduce the *area* of the inductive energy storage unit for the voltage regulator. The coupled inductor can store energy as not only *self-inductance* but also *mutual* or *magnetizing* inductance. Figure 2.18 gives a simple schematic of an ideal coupled inductor that is used in a buck regulator.

In the figure, the main parameters of the coupled inductor are the coupling factor K, the leakage inductance L_{σ} or L_l (the leakage inductance), and L_M (the magnetizing inductance). L_1 and L_2 are the two *self-inductances* of the inductor. Figure 2.19 shows of how the inductors can be coupled physically. There are many other combinations and geometries in which a coupled inductor may be constructed. This figure gives the simplest illustration to help the reader understand the basic construction. In the figure, A_c is the area of one of the inductor regions—in this drawing, the

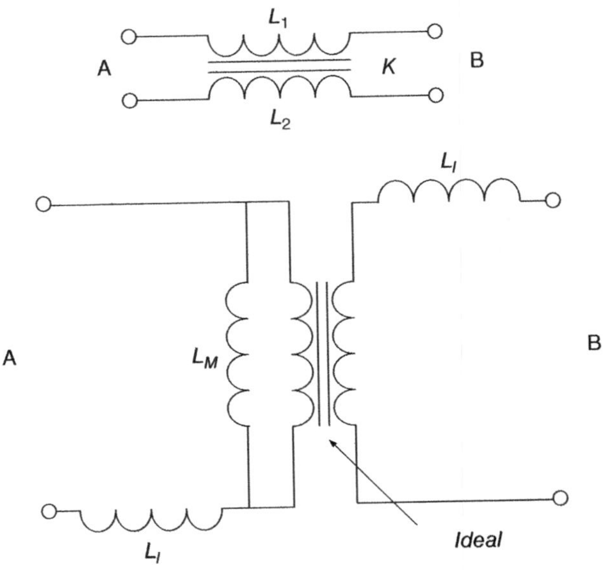

Figure 2.18 Ideal coupled inductor circuit schematic

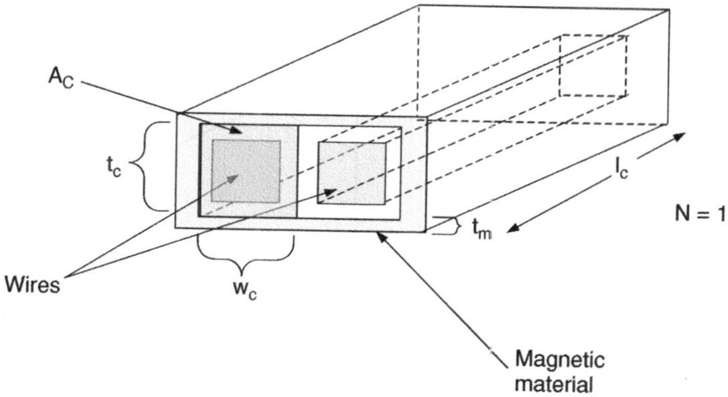

Figure 2.19 Simple isometric of coupled inductor

turns ratio is 1. There have been numerous papers written on converters that utilize coupled inductors [6]. The reader is referred to a few of these references at the end of the chapter.

When estimating the losses in coupled inductors, it is important to consider the AC and DC losses as described in the previous section. The main differences will be in the losses through the magnetizing inductance (L_m), which should be negligible for most ferrite-based structures. Thus, in general, the equations from the previous section can be used to compute the losses, and this means treating the inductors as two independent structures. Since the coupled inductor is used in buck converters as well, the pair of inductors shown as a *transformer* in Figure 2.18 will be used as a *two-phase* pair where the nodes to the bridge phases will be connected to one side of the adjacent *A* and *B* connections. Coupled inductor structures can be also used in multi-phase converters to reduce the platform area. Multi-phase converters have the benefit of reducing ripple and current at each phase while achieving the same power output (coupled or not). These converters are discussed briefly in the next section.

2.2.9 Multi-phase Buck Converters

The multi-phase buck converter is today one of the most popular versions of the buck regulator. As currents have increased for high-performance devices, the need to deliver high-quality highly efficient power has been a challenge. This is because the single-phase buck regulator is limited by the current passing through the inductor, which must increase along with the current for the silicon it is delivering in order to maintain the required power to the load.

Multi-phase buck converters further differ from single-phase converters in that they reduce the output ripple current. In turn, this helps reduce the physical size of the inductor, and thus cost. Ripple reduction is caused by the shifting in the control of the drive at each phase (e.g., in a two-phase design, one phase can be run 180° from the other), so that the ripple currents are made to cancel to some

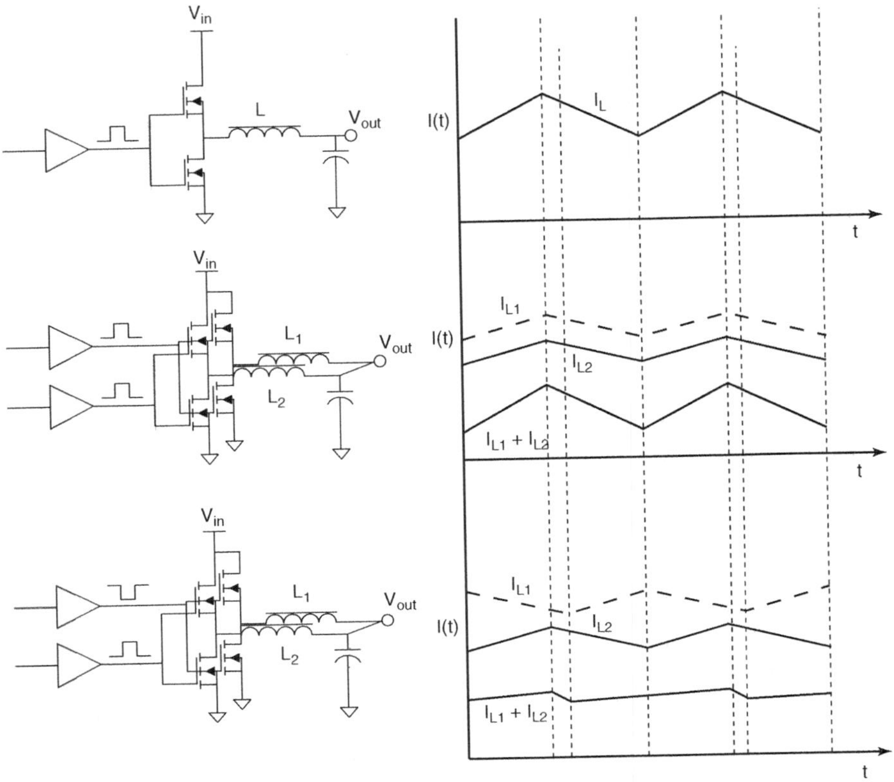

Figure 2.20 Single-phase, two-phase, and two-phase (polar operation)

extent at the output of each inductor—which is tied to the same node. Figure 2.20 shows the topology differences between single-phase and multi-phase (two-phase) converters. A single-phase converter has to deliver the *entire* energy to the load, whereas a two-phase system converter splits the load in two. When the phases are operated 180° out of phase (polar operation), much of the ripple current is canceled. This results in less ripple voltage at the output (again, all else being equal).

The phase relationships between *n* phases (greater than 2) is determined by the *mode* of operation of the multi-phase converter. Often the phase imbalance is *forced* by the controller to determine the best phase angle for the minimal ripple. For example, a four-phase converter would typically operate all phases 90° out of phase with each other. This would result in a ripple that is four times the frequency of the switching frequency for one phase. The advantage, of course, is that the ripple voltage at the output node of the converter is lower. Thus, many designers opt for reducing the amount of capacitance when using a multi-phase design, which reduces BOM cost, area, and can improve reliability. Unfortunately, this will also create a ripple that is higher in frequency, and sometimes makes decoupling a greater challenge. Coupled inductor systems can also be used in a multi-phase configuration, as discussed in the previous section. This has the added advantage of not only reducing the ripple—due to the two phases—but also shrinking the physical size of the inductor solution, since the energy is now stored *between* the inductors. The next section introduces a less common buck topology but one that is gaining interest on some systems—the tapped-inductor buck converter.

2.2.10 The Tapped-Inductor Buck Converter

Tapped-inductor topology has been in use for several years and in many applications. Over the last few years, designers have been finding new ways to utlilize this topology for computer platforms as well [7]. That tapped-inductor offers some more attractive features compared to the standard buck regulator and can additionally be used in a multi-phase configuration. The main advantage is that it extends the duty cycle of a standard buck regulator. When the duty cycle is small (e.g., with a 12 V_{in} 1.5 V_{out} buck) the losses can be high, due to the larger inductor ripple currents, that keep on increasing both switching and conduction losses in the power FETs and the inductors. By increasing this duty cycle, all else being equal, these losses can be reduced. The tapped-inductor buck regulator, for a *single-phase* design, is shown in Figure 2.21.

The tapped-inductor buck, or TI buck for short, has two windings, one on the primary or V_{in} side, and one at the output as in a normal buck regulator. The topology shown in the figure is a variant on the classic TI buck. The difference here is that the upper switch is connected on the bottom side of the top inductor. This is done to allow for ease in driving the upper FET, which is usually an NMOS transistor. When the upper FET turns on, current flows in both inductors and to the output V_{out}. When the upper FET is turned off, the voltage across both inductors reverses and current continues to conduct into the output inductor from the top inductor. The

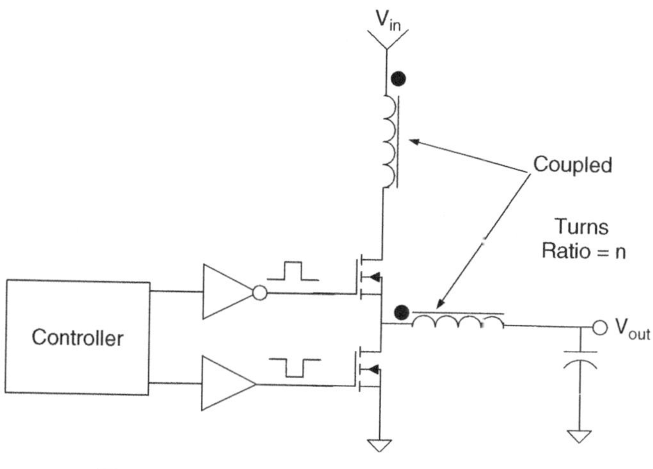

Figure 2.21 Tapped-inductor buck converter

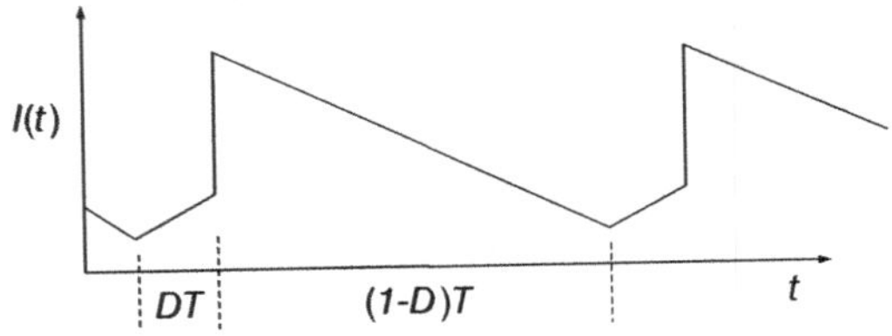

Figure 2.22 Current waveform for tapped-inductor buck regulator

relationship of the duty-cycle, turns ratio, and voltages is expressed as

$$\frac{V_{Out}}{V_{In}} = \frac{D}{D + n(1 - D)} \tag{2.53}$$

The higher the turns ratio is, the *larger* is the duty-cycle. One big problem with this design is that a large voltage spike occurs across the upper FET when it turns off. The typical current waveform is shown in Figure 2.22.

In the tapped-inductor buck topology, the current is discontinuous to some extent as shown in the figure. This is caused by the coupling with the primary side inductor. The large current rise between the duty and commutation cycle can cause a large voltage spike as well. This large voltage can damage and create reliability issues in the upper device, due in part to the reversal of voltage across the upper inductor. However, there have been some novel methods tried to control this damage. Some designers have achieved relatively high efficiencies with this topology [8–12]. By controlling the voltage spike with either an active or a passive network, the large spike can be sufficiently mitigated to reduce the potential for damage. Additional winding on the main inductor has been tried, but this can increase the design cost well beyond that of a conventional buck regulator. Nevertheless, often this trade-off is worth it, particularly in designs that are constrained by thermals that necessitate high efficiency, low energy loss.

As advancements in devices and passive technologies progress, and as duty cycles shrink, the TI buck may become a good option for VR designers.

2.3 LAYOUT AND NOISE CONSIDERATIONS

What typically separates a *good* converter design from a *poor* one, with the exception of efficiency, is the ripple and noise that it generates. Most designs today are *highly integrated*, meaning that the functions in the silicon portions are many and sophisticated as compared with VR designs from years past. This level of integration usually allows for good noise suppression if the internal silicon has good decoupling and is designed adequately as well. Over the years, converter designs have *split* the functions into the power train and the control, with only a few components external for decoupling and bias power. Thus, today we see the FETs, drivers, and filters distinct from the controllers in the converter. The advancements in silicon has helped facilitate this fundamental separation. The controllers contain the low noise, high gain amplifiers for feedback and sensing, which need to be well decoupled, while the power train contains the high-power and high-current switching components, which typically generate a plethora of large noise signals in the frequency domain. Thus, the control signals that are used to drive these devices are often of lower voltage and much lower current as compared to the power train.

Though the designer must also be careful in the layout of the controller and the power FETs to prevent noise from coupling back into the controller amplifiers, care must also be taken so as to not generate too much noise into the silicon load that the converter is powering. This is where the power integrity engineer usually spends a good portion of time analyzing and measuring the noise in the system. Important to delivering high-quality power, is additionally the analysis of such issues affecting the PDN. Noise analysis is discussed in more detail in Chapter 6. In this section the focus will be on the different types of generic noise generation and how they can be prevented in the basic buck converter design and layout.

The noise is generated from the converter is usually one of four main types: *low-frequency ripple*, *switching* from the power FETs and drivers, *ring-back* due to parasitics in the layout path of the power train and filter FETs, and/or *ground and power noise* (mostly due to poor layout design of the power train). Notice that there was no mention of noise generated from the controller itself. Though the edge rates of the control signals are relatively fast (tens of nanoseconds or less), the energy content is typically low, and with good plane decoupling, a designer can usually prevent this noise from becoming an issue. Table 2.5 lists the noise sources and the ways they are often mitigated.

TABLE 2.5 Noise sources and possible mitigation techniques

Sources of Noise	Mitigation Method(s)
Low-frequency ripple	*Use multiple phases out of phase, use more low-frequency caps with lower ESR, lower impedance of power/ground plane pairs*
FET switching noise	*Lower slew rate of FETs, shift dead-time of upper and lower FETs, use soft-switch topology, use HF decoupling at noise of FETs, reduce parasitic connections to main filter, lower loop impedance between output of main filter and power/ground connections of main FETs/drivers*
Parasitic ring-back	*Use lower impedance path between main power train filter and FETs, use low inductance loop at same connection, reduce input parasitics from input filter to main power FETs/drivers*
Power/ground plane noise	*Lower impedance (typically lower inductance) between power and ground planes, place HF decoupling caps across power and ground planes in regions nearest filter output of main power train filter*

Many noise issues are hard to track down, and not all of the methods in Table 2.5 may work if employed alone. Usually it is a combination of techniques that will reduce the noise to an acceptable level. Moreover, as mentioned earlier in the chapter, the power converter designer is often constrained by numerous boundary conditions (area, cost, etc.) that can make some of these applications a challenge.

Another important area for the PI engineer to consider is the *physical structure* of the converter and its impact on the system-level issues. This is basically the placement of the components relative to the load and the PDN decoupling and layout. The PI engineer must indeed ensure that the coupling of the noise is mitigated at every level, particularly as the power is being delivered to a high-performance, low-voltage silicon device. The placement of components also has an effect on the power loss of the converter in a number of ways. Remember, understanding the power loss in the platform regulator depends on understanding the losses in the *components*. Much was made of this in the previous sections—and in particular, concerning the power FETs and inductors. Because most switching regulators have significant AC and DC currents, knowing the effects from a layout perspective (parasitics) can greatly help in an efficiency estimation. Dependence on a circuit schematic is insufficient to determine the overall power loss in a converter. This is because the circuit schematic does not show the interconnect model losses. Thus, it is usually a productive exercise to sit down with the board layout engineer and examine some of the key traces and then estimate the resistance of these sections to determine the losses at the interconnects. It has been shown that there are impacts from both inductors and the switches in a buck regulator, for example. There are also losses due to the capacitive storage elements. This is discussed in Chapter 4, where capacitor technologies are reviewed. The basic equations for estimating interconnect parasitics and other electromagnetic structures are presented in the next chapter.

2.4 SUMMARY

This chapter gave an introductory overview of computer platform conversion. The discussion began by describing centralized and distributed systems. Smaller systems tend to be centralized, whereas larger ones are distributed. Next, a brief introduction was provided to the linear regulator and the buck platform converter types. Some simple examples were given on how to compute losses for both the linear regulator and the buck converters, and it was shown that for higher power applications, the buck regulator is usually a better fit. A discussion of the LC filter showed how this filter could affect ripple. The basic components of the buck converter were introduced to familiarize the reader with their use before later chapters embark on modeling their purposes. The power MOSFETs, inductor, and controller were briefly discussed. Filter capacitance will be covered more fully in Chapter 4. A few different buck topologies were described along with basics on the coupled inductor, which is gaining in popularity due to its small form. Multi-phase converters and the tapped-inductor converter were examined as complements to the classic single-phase buck regulator. Finally, a short discussion on noise and layout considerations was given; more on this subject are provided in later chapters.

REFERENCES

1. Kassakian, J. G., Schlecht, M. F., and Verghese, G. C. *Principals of Power Electronics*. Prentice-Hall, 1991.
2. Divakar BP, Ioinovici A. Zero-voltage-transition converter with low conduction losses operating at constant switching frequency. *Power Electronics Specialists Conference*. Baveno, 1996.
3. Lee, Y. S., Wang, S. J. and Hui, S. Y. R. Modeling, analysis, and application of buck converters in discontinuous-input-voltage mode operation. 2, *IEEE Trans. Power Electronics*, Vol. 12, 1997.
4. Baliga, B. J. *Advanced Power MOSFET Concepts*. Springer; 2010.
5. Erickson, R. W., and Maksimovic, D. *Fundamentals of Power Electronics, 2nd ed.*, 2000.
6. Zumel, P., Chen, D., Liu, C-W, Huang, K, Tseng, E., Tai, B. Tight Magnetic Coupling in multiphase interleaved converters based on simple transformers, *IEEE Applied Power Electronics Conference*,Vol. 1, 2005.
7. Lee, M., et al. Comparisons of three control schemes for adaptive voltage position (AVP) droop for VRMs applications. Power Electronics and Motion Control Conference, 2006.
8. Xu, P., Wei, J., Yao, K., Meng, Y., Lee, F. C. Investigation of candidate topologies for 12V VRM. APEC, 2002.
9. Harris, P., and DiBene II, J.T. *Integrated Magnetic Buck Converter with Magnetically Coupled Synchronously Rectified Mosfet Gate Drive*, U.S. Patent # 6,754,086– June, 2004.
10. DiBene II, J. T., Morrow, P. R., Park, C-M, Koertzen, H. W., Zou, P., Thenus, F., Li, X., Montgomery S. W., Stanford E., Fite, R., Fischer, P. A 400 Amp fully integrated silicon voltage regulator with in-die magnetically coupled embedded inductors. IEEE APEC Conference 2010.
11. Vafaie, M. H., Adib, E., Farzanehfard, H. A self powered gate drive circuit for tapped inductor buck converter. PEDSTC, 2012.
12. Kingston, J. Application of a passive lossless snubber to a tapped inductor buck DC/DC converter. *IEEE PEMD*, 2002.

PROBLEMS

2.1 Repeat Example 2.1, except now assume that the output voltage is 1.2 V and the converter efficiency is only 78%. What is the total resistance and current in the interconnect if the temperature is now 96°C instead of 80°C?

2.2 How much current can a linear regulator design safely provide if the temperature is held at the limits of the operation of the device? Use the same data as in Example 2.2.

2.3 Compute the maximum current for a buck regulator—ignoring extraneous losses—if the input voltage is 12 V and the inductor is 0.47 μH. Assume that the duty cycle is 6.5% and the switching frequency is 1.4 MHz. Also assume that the rise and fall times for the switches are negligible, and that the regulator operates in full conduction mode where the minimum current point is 20% of the maximum point. What is the *average* current in this design for the same parameters? Assume perfect triangles for the computation of the current. What is the output power?

2.4 Compute the power loss in the *upper* switch in Example 2.4.

2.5 Derive the RMS current in equation (2.43).

2.6 Compute the power loss using the Steinmetz parameters for an inductor given these values for k, α, and β (along with f and B):

PARAMETER	VALUE
k	1.56 e^{-8}
α	0.19
f_{sw}	500 KHz
β	2.19
B	750

Assume that the *true* frequency needed in the calculation comes from the following formula:

$$f_{(eff)} = \frac{f_{sw}}{2\pi(d - d^2)} \tag{2.54}$$

where d is the duty cycle. Assume that the duty cycle is the same as in Problem 2.3.

2.7 Determine the inverse Laplace transform for equation (2.37).

3

REVIEW OF ELECTROMAGNETIC FIELD AND CIRCUIT REPRESENTATIONS

Before analyzing the power distribution networks that connect the power converters to the silicon (and are distributed within the silicon), it is usually a good idea to review the governing field equations and the circuit representations, as are approximated from those fields. All of the passive elements that are used for *first principles analyses* and SPICE modeling come from these fundamental electromagnetic equations, called *Maxwell's equations*.

Virtually all of the values in the distribution networks analyses are approximations. However, the approximations are usually sufficient to enable one to gain insight to the problems at hand. It is critical, though, that the warnings presented in Chapter 1 be heeded, since it is too easy to fall into the trap of believing that the approximations yield the *exact* result or even highly accurate results. Nonetheless, approximations are excellent starting points for bounding difficult problems prior to embarking on a serious numerical study. Thus, a discussion of where these approximations tend to fall short and where they are usually "good enough" for a first pass analysis will be presented here. Other ways in which extractions for passive elements may be obtained are through numerical programs, or *field solvers*. There are many good extraction tools that exist today, and texts that explain how they work; a few examples are given at the end of the chapter [1,2]. In most complex problems, these tools will yield reasonable approximations that can be used for the final system simulations. Unfortunately, even with good tools, many engineers still rely too heavily on extraction programs to generate the data for them.

Fundamentals of Power Integrity for Computer Platforms and Systems. First Edition. J. Ted DiBene II.

But these tools are only as good as the data and *assumptions* that go into them, and these tools still only compute estimations to the actual field data. Additionally, too often there is an urge to neglect basic field behavior when applying—for example, in deciding where to place the ground returns in the model or adjacent circuit coupling. Even then, it is rare that accurate extraction of passive elements (e.g., inductance) is better than 15% to 20%. Thus, there must always be a check of the results against known validated data and/or similar extractions from reliable sources. Often it is wise to start with basic *hand* calculations prior to starting the more detailed tool extraction process to ensure that simulations are yielding reasonable results. For the power integrity engineer who is solving a circuit level problem, the process of understanding the fundamental field behavior—before embarking on a rigorous tool extraction—is a necessary step. This chapter will begin with an introduction to the field and circuit equations that are the basis for vector calculus. A review of some of the basic equations in electrostatics will be presented along with a discussion on how to utilize the results to solve simple *first principles* problems in closed form. Then there will be discussion of some of the more common geometries found in power distribution analyses. The chapter will start with a very brief overview of vector analysis. It is assumed here that the reader has a basic understanding of vector calculus and electromagnetics. Thus, this chapter is intended to serve as a *refresher* of those concepts. There are many excellent references [3–7] that discuss electromagnetics and circuit extraction in detail. This book will only touch the surface of this broad and critical subject. The reader is encouraged to review these references and other sources for a more in-depth study.

3.1 VECTORS AND SCALARS

Maxwell's equations are basic to understanding electromagnetic fields and related circuit behavior for virtually every power integrity problem that an engineer will encounter. In particular, this field behavior is represented by *vectors*. Remember, a vector has both magnitude and direction associated with it. For example, for a stick whose tip is pushed against a solid flat surface, the force exerted on this stick is represented by a vector, as shown in Figure 3.1.

Since the field has both *magnitude* and *direction*, the net force is given by $\mathbf{F} = \mathbf{F}_x + \mathbf{F}_y$. Similarly, for a voltage applied across two plates with a battery, a vector field $\mathbf{E}_y$ will be created due to the charge Q that has built up in that direction as shown in the figure. A vector field is typically represented using boldface characters. A vector may also be multiplied by a scalar λ to change its magnitude. A few examples of how to add, subtract, and multiply vectors are shown in equations (3.1) through (3.4).

$$\mathbf{B} + \mathbf{C} = \mathbf{C} + \mathbf{B} \tag{3.1}$$

$$\lambda(\mathbf{B} + \mathbf{C}) = \lambda\mathbf{B} + \lambda\mathbf{C} \tag{3.2}$$

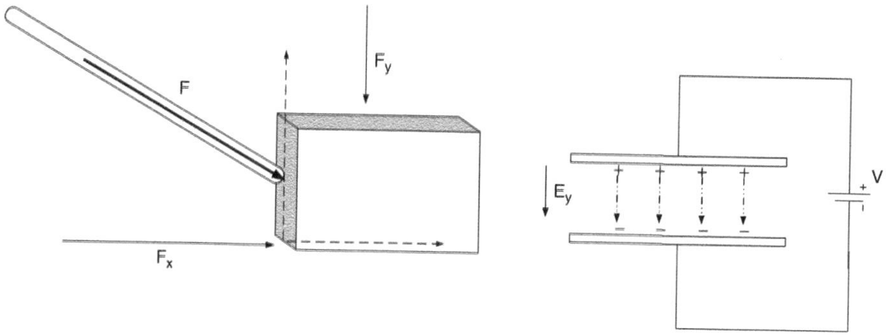

Figure 3.1 Examples of vector fields

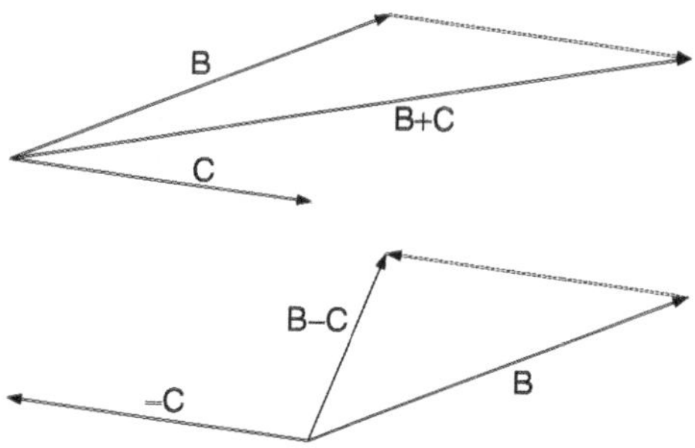

Figure 3.2 Vector addition and subtraction

$$\mathbf{A} + (\mathbf{B} + \mathbf{C}) = (\mathbf{A} + \mathbf{B}) + \mathbf{C} \tag{3.3}$$

$$\mathbf{B} - \mathbf{C} = \mathbf{B} + (-\mathbf{C}) \tag{3.4}$$

Adding vectors can be done mathematically or graphically as shown in Figure 3.2, for simple vectors in two or three dimensions. When adding one vector to another graphically, one vector is placed in space and the other is placed at its tip by drawing a line connecting the tail of the first vector to the tip of the other as shown in Figure 3.2. Note that in both cases, vector **C** has been shifted (dashed line) to align to the tip of vector **B**. A discussion of vectors in *coordinate systems* is presented in the next section.

3.1.1 Coordinate Systems

Depending on the problem, it is often easier to represent them problems in the coordinate systems, which will simplify the computation and/or solution to the problem. For complex numerical analyses, a non-orthogonal coordinate system may be computationally more efficient than an orthogonal system. In this text, the focus will be on the three most common coordinate systems, which are all orthogonal: *cartesian*, *cylindrical*, and *spherical*. For many of the geometries that will be encountered, *cartesian* and *cylindrical* usually suffice to describe most of the problems encountered in power integrity. The *spherical* coordinate system descriptions will be described in the appendix.

The *cartesian* coordinate system in three dimensions is shown in Figure 3.3. Many problems are described and easily solved in two dimensions. Three-dimensional coordinates are simply an extension of the two-dimensional system. For example, a vector in this system may be described by breaking the vector into its three orthogonal coordinates and their coefficients as in the following equation:

$$\mathbf{A} = A_x \boldsymbol{u}_x + A_y \boldsymbol{u}_y + A_z \boldsymbol{u}_z \tag{3.5}$$

where $\boldsymbol{u}_n$ is the unit vector for that dimension and A_n are the coefficients, respectively. The coordinates (x, y, z), are the orthogonal vector directions for this coordinate system and the unit vector $\boldsymbol{u}_n$ is defined here at any generic point in the system. The coefficients A_n are the generic coefficients defined for that vector correlating to that particular unit vector. If it is required to define a vector in this space at a particular point, a point can be chosen in space and represented by that vector as

$$\mathbf{A} = A_{x1} \boldsymbol{u}_{x1} + A_{y1} \boldsymbol{u}_{y1} + A_{z1} \boldsymbol{u}_{z1} \tag{3.6}$$

where vector $\mathbf{A}$ is defined at point $P(x_1, y_1, z_1)$ in space. For $P(0, 0, 0)$ this point is the origin in Figure 3.3. Additionally, the magnitude is found using

$$|\mathbf{A}| = \sqrt{A_{x1}{}^2 + A_{y1}{}^2 + A_{z1}{}^2} \tag{3.7}$$

A unit vector in the direction of vector $\mathbf{A}$ is

$$\hat{\mathbf{a}} = \frac{\mathbf{A}}{|\mathbf{A}|} \tag{3.8}$$

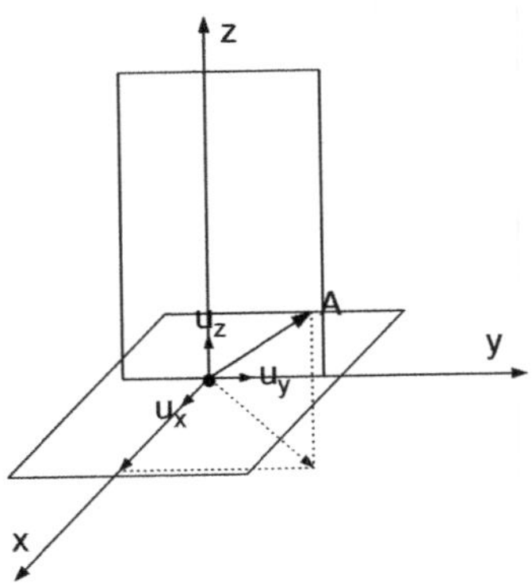

Figure 3.3 Illustration of a vector **A** in cartesian coordinates

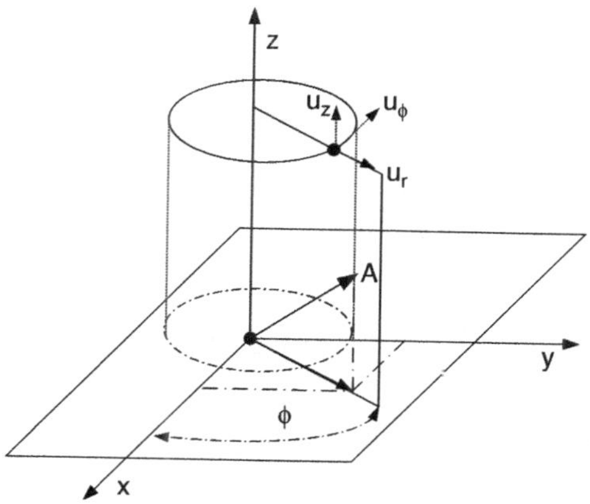

Figure 3.4 Cylindrical coordinate system

That same vector can be represented in *cylindrical* coordinates (Figure 3.4). Many problems lend themselves to a simpler setup and solution when the proper coordinate system is chosen.

Vector **A**, in cylindrical coordinates, becomes

$$\mathbf{A} = A_r \boldsymbol{u}_r + A_\phi \boldsymbol{u}_\phi + A_z \boldsymbol{u}_z \tag{3.9}$$

where conversion between coordinates is accomplished using

$$x = r \cos \phi, \quad y = r \sin \phi \tag{3.10}$$

Example 3.1 Add the two vectors below, find the magnitude, and represent the solution in cylindrical coordinates:

$$\mathbf{A} = 4\boldsymbol{u}_x + 3\boldsymbol{u}_y + 5\boldsymbol{u}_z \tag{3.11}$$

$$\mathbf{B} = 7\boldsymbol{u}_x + 2\,\boldsymbol{u}_y + 6\boldsymbol{u}_z \tag{3.12}$$

SOLUTION:

$$\mathbf{A} + \mathbf{B} = \mathbf{C} = (4+7)\boldsymbol{u}_x + (3+2)\boldsymbol{u}_y + (5+6)\boldsymbol{u}_z = 11\boldsymbol{u}_x + 5\boldsymbol{u}_y + 11\boldsymbol{u}_z \tag{3.13}$$

$$|\mathbf{C}| = \sqrt{\mathbf{11}^2 + \mathbf{5}^2 + \mathbf{11}^2} = \sqrt{267} = 16.34 \tag{3.14}$$

$$\mathbf{r} = \sqrt{x^2 + y^2} = \sqrt{136} = 11.66 \tag{3.15}$$

The solution now needs to be translated into cylindrical coordinates. The magnitude of each coefficient may be found from the previous translation equations:

$$\mathbf{C} = C_r \boldsymbol{u}_r + C_\phi \boldsymbol{u}_\phi + C_z \boldsymbol{u}_z \tag{3.16}$$

where conversions from equation (3.10) are now used:

$$C_r = \sqrt{136}\ \cos\left(\tan^{-1}\left(\frac{5}{11}\right)\right) = 11.7 \times \cos(24.4) = \widetilde{8.7} \tag{3.17}$$

$$C_\phi = \sqrt{136}\ \sin\left(\tan^{-1}\left(\frac{5}{11}\right)\right) = 11.7 \times \sin(24.4) = \widetilde{4.8} \tag{3.18}$$

and C_z will be the same as in the rectangular coordinates.

3.1.2 Vector Operations and Vector Calculus

Some basic formulas are reviewed here and used to manipulate and describe the fields and equations that will be introduced in the next section. These are some of the equations and mathematical tools that will be used in solving the basic power integrity problems in subsequent chapters. Many involve the setting up vector equations whose solutions are obtained through *first principles* or numerical analyses. For the most part, numerical solvers do not reveal these equations to users, and thus it is up to the engineer to understand the limits of the tool and input the data correctly to get the most accurate result. The engineer who does not understand the convergence or accuracy limitations in a solver could end up with faulty results [1,2]. Moreover, a good foundation in the fundamentals of the equations that are used in such solvers helps the engineer in arriving to a more precise result when using such tools. The reader who is not familiar with the formulas in this chapter is urged to review some of the references at the end of the chapter for a more comprehensive study, since this section, once again, is only a very brief introduction.

Two fundamental vector operations will be defined here; the *dot* and *cross* (or *vector*) product. The dot product is the multiplication of the scalar quantities of the coordinate coefficients of two vectors and its result is a scalar:

$$\mathbf{A} \cdot \mathbf{B} = |\mathbf{A}||\mathbf{B}|\ \cos\theta \tag{3.19}$$

where θ is the angle between **A** and **B** in equation (3.19) The dot product of two vectors may also be found by the more common relation

$$\begin{aligned}\mathbf{A} \cdot \mathbf{B} &= (A_x \boldsymbol{u}_x + A_y \boldsymbol{u}_y + A_z \boldsymbol{u}_z) \cdot (B_x \boldsymbol{u}_x + B_y \boldsymbol{u}_y + B_z \boldsymbol{u}_z) \\ &= (A_x \cdot B_x + A_y \cdot B_y + A_z \cdot B_z)\end{aligned} \tag{3.20}$$

Conversely, the cross product result is a *vector* quantity that is orthogonal to the plane in which the two vectors reside:

$$\mathbf{A} \times \mathbf{B} = |\mathbf{A}||\mathbf{B}|\sin\theta \boldsymbol{u}_n \tag{3.21}$$

where $\boldsymbol{u}_n$ is the unit vector normal to **A** and **B**. The more conventional calculation may be found from the determinant of the matrix

$$\mathbf{A} \times \mathbf{B} = \begin{vmatrix} \boldsymbol{u}_x & \boldsymbol{u}_y & \boldsymbol{u}_z \\ A_x & A_y & A_z \\ B_x & B_y & B_z \end{vmatrix} = \boldsymbol{u}_x(A_yB_z - A_zB_y) - \boldsymbol{u}_y(A_xB_z - A_zB_x)$$
$$+ \boldsymbol{u}_z(A_xB_y - A_yB_x) \tag{3.22}$$

The dot and vector products are used extensively when it is required to manipulate Maxwell's equations. The *del* operator (an upside-down triangle) is used when the gradient of a scalar or vector field is needed. The gradient of a scalar field (e.g., the temperature gradient in the ocean) or a vector field (e.g., the flow of heat across a piece of metal from one end to the other) can be described using this operator. The *del* operator, in cartesian coordinates, is defined in equation (3.23),

$$\nabla = \frac{\partial}{\partial x}\boldsymbol{u}_x + \frac{\partial}{\partial y}\boldsymbol{u}_y + \frac{\partial}{\partial z}\boldsymbol{u}_z \tag{3.23}$$

The result of applying the del operator to a function, for example, is a vector. To find the gradient of field **f**, the definition used is

$$\textbf{grad f} = \nabla f = \frac{\partial f}{\partial x}\boldsymbol{u}_x + \frac{\partial f}{\partial y}\boldsymbol{u}_y + \frac{\partial f}{\partial z}\boldsymbol{u}_z \tag{3.24}$$

which results in a vector. The divergence of a field is found by taking the dot product of a field,

$$\nabla \cdot \mathbf{F} = \frac{\partial}{\partial x}\boldsymbol{u}_x \cdot F_x\boldsymbol{u}_x + \frac{\partial}{\partial y}\boldsymbol{u}_y \cdot F_y\boldsymbol{u}_y + \frac{\partial}{\partial z}\boldsymbol{u}_z \cdot F_z\boldsymbol{u}_z$$
$$= \frac{\partial F_x}{\partial x} + \frac{\partial F_y}{\partial y} + \frac{\partial F_z}{\partial z} \tag{3.25}$$

which results in a *scalar*. The curl operation is found by applying the cross product using the *del* operator to a vector field which results in a vector. Applying the equations above for the cross product and *del* operator yields

$$\nabla \times \mathbf{F} = \begin{vmatrix} \boldsymbol{u}_x & \boldsymbol{u}_y & \boldsymbol{u}_z \\ \dfrac{\partial}{\partial x} & \dfrac{\partial}{\partial y} & \dfrac{\partial}{\partial z} \\ F_x & F_y & F_z \end{vmatrix} = \boldsymbol{u}_x\left(\frac{\partial F_z}{\partial y} - \frac{\partial F_y}{\partial z}\right) - \boldsymbol{u}_y\left(\frac{\partial F_x}{\partial z} - \frac{\partial F_x}{\partial x}\right)$$
$$+ \boldsymbol{u}_z\left(\frac{\partial F_y}{\partial x} - \frac{\partial F_x}{\partial y}\right) \tag{3.26}$$

Equation (3.26) is a useful for setting up problems to be solved using numerical methods. All these equations will be used throughout this chapter to illustrate how to solve various electromagnetic problems.

Another powerful formula that is encountered in electromagnetics is the *divergence theorem*, which states: the total flow through a surface is equal to the sources (minus the sinks) within that surface:

$$\oiint \vec{\mathbf{F}} \cdot \mathbf{d}\vec{\mathbf{S}} = \iiint \nabla \cdot \vec{\mathbf{F}} dV \qquad (3.27)$$

The divergence theorem allows the conversion from a *volume* integral to a *surface* integral, which often simplifies the analysis for many problems. The divergence theorem of equation (3.27) will be used shortly in the discussion of electrostatics. Another very useful formula is *Stoke's theorem*, which relates the surface integral of the curl of a field to an equivalent closed line integral of the field over that boundary:

$$\iint \nabla \times \vec{\mathbf{F}} \cdot \mathbf{d}\vec{\mathbf{S}} = \oint \vec{\mathbf{F}} d\vec{\mathbf{r}} \qquad (3.28)$$

Stoke's theorem and the divergence theorem will be used throughout these discussions to help simplify a number of the analyses that will be encountered. A number of identities and formulas are also given in the Appendix for the reader to purview.

3.2 STATIC FIELDS

The first section of this chapter reviewed of some of the basic tools for vector calculus. Now some of those tools will be used to introduce electromagnetic field concepts. The first objective will be to introduce those fields in relatively simple terms before introducing Maxwell's equations. Thus, the focus here will be on static field concepts. There are two types of fields that are necessary when dealing with power integrity problems—*electrostatic* and *magneto-static*. Electrostatic problems deal with *charge* accumulation between one object and another. Magneto-static problems deal with *magnetic fields* due to currents within a structure. The discussion will start with the former. Static field concepts are critical to understanding power integrity problems, since most of these problems will deal with frequencies that are well approximated by static field concepts. For the static equations herein, the vector symbol will be omitted and the use of ***bold italics*** will be used to represent the static field vectors.

3.2.1 Electrostatics

Charge is one of the fundamental sources of energy that people encounter daily. It allows individuals to communicate, mobilize, and advance. The measure of charge is the Coulomb. The charge density and distribution are fundamental to understanding how to manage power integrity problems. As an example, the charge contained in a capacitor is used to determine the filter characteristics of a distribution problem.

However, typically, large quantities of charge rather than individual charges are encountered in nature and thus the interest for engineers is mainly in the total *charge* or *charge density*. For example, engineers typically compute the capacitance of a capacitor by knowing the total charge and voltage across it:

$$Q = CV \quad \text{or} \quad C = \frac{Q}{V} \tag{3.29}$$

The total charge is determined by finding the charge density across the plate and then integrating, or summing, over that region. This is illustrated in the following example.

Example 3.2 A capacitor (Figure 3.5) with a charge density across its plate of

$$\rho(x,\ y) = 2x + 3y\ \text{C/mm}^2 \tag{3.30}$$

is charged to 1 V. If the capacitor is 5 × 5 mm in size, compute the total charge on the capacitor and then the total capacitance.

SOLUTION: The total charge is found by integrating the charge from equation (3.30) over the area (note that it is multiplied by 1e-6 below because the boundary segment, or length of each side of the capacitor, in each integral is in mm):

$$Q = \int_0^5 \int_0^5 (2x + 3y)dxdy = (25 + 1.5 \times 25) \times 1e - 6 = 62.5e - 6C = C \times 1$$

$$\text{or} \quad C = 62.5\ \mu\text{F} \tag{3.31}$$

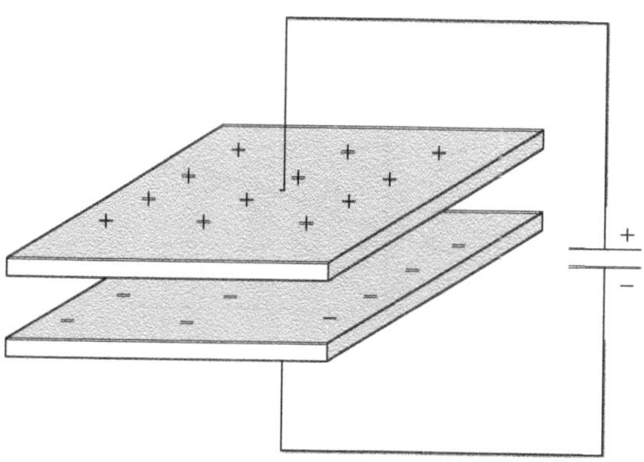

Figure 3.5 Parallel plate capacitor

For many PI problems, simple capacitance computations are useful in studies of the elements within the distribution path. Capacitance concepts will be discussed later in the chapter.

The preceding example shows at a macro level how to compute the total charge on a capacitor given a certain charge density. But, what about the actual charges? At the electron level, it is simple to compute the force between two charges given their distance and the medium (called the *dielectric*, which is air in this case) separating them.

Suppose, for the moment, that the two charges are in space as shown in Figure 3.6. The force exerted on q_2 by q_1 is computed using *Coulombs* law:

$$\vec{\mathbf{F}} = \frac{q_1 q_2}{4\pi\epsilon_0 r^2}\vec{\mathbf{a}}_r = \frac{q_1 q_2(\boldsymbol{r}_1 - \boldsymbol{r}_2)}{4\pi\epsilon_0 |\boldsymbol{r}_1 - \boldsymbol{r}_2|^3} N \tag{3.32}$$

where vector $\boldsymbol{R}$ is equal to $\boldsymbol{r}_1 - \boldsymbol{r}_2$ as shown in Figure 3.6. The unit of force here is the *Newton*, a familiar unit from the study of *mechanics*. The unit of charge is the Coulomb, or $1\ C = 1 \times 10^{-18}$ EV or 1 electron volt. Note that the force is a *vector* quantity and the force that is exerted on q_1 by q_2 is in the direction of the unit vector $\boldsymbol{a}_r$ as shown. The constant ϵ_0, is called the permittivity of free space and is given the value

$$\epsilon_0 = 8.854 \times 10^{-12}\ \text{F/m} \tag{3.33}$$

If the fields are in dielectrics other than free space, the total permittivity is determined by the permittivity constant for that dielectric multiplied by ϵ_0,

$$\epsilon = \epsilon_r \epsilon_0 \tag{3.34}$$

If there is more than one charge in space, the total charge exerted onto charge q_1 will be the sum of all the forces on that charge, as shown in Figure 3.7.

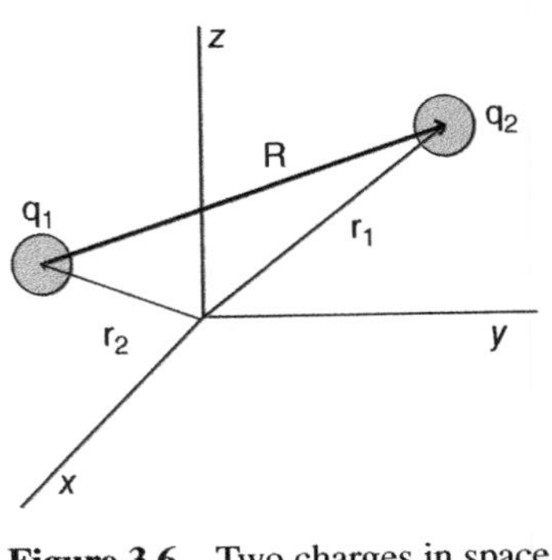

Figure 3.6 Two charges in space

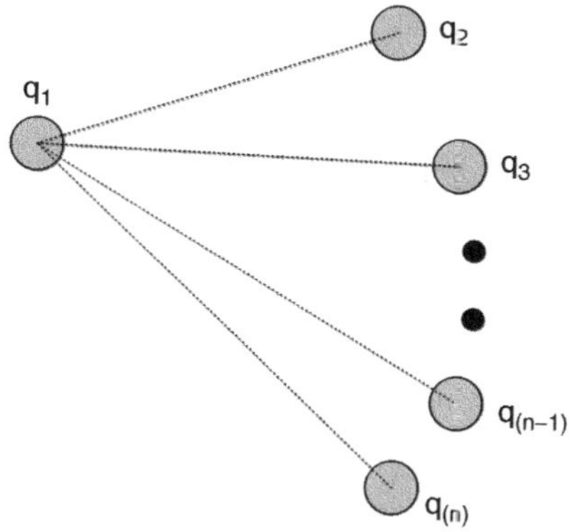

Figure 3.7 Charges in space

Since each charge is separated from q_1 by a specific distance (which is assumed to be different in this case), the forces are summed up *vectorially*, based on each charge position relative to q_1 to obtain the total force acting on q_1:

$$\textit{Total force} = \sum \vec{\mathrm{F}} = \sum_{i+1}^{n} \frac{q_1 q_i (\boldsymbol{r} - \boldsymbol{r}_{i+1})}{4\pi\epsilon_0 |\boldsymbol{r} - \boldsymbol{r}_{i+1}|^3} N \tag{3.35}$$

Notice that $\boldsymbol{r}$ is the vector from the origin to charge q_1 as in Figure 3.6, while $\boldsymbol{R}$ is the resultant vector from subtracting each vector, $\boldsymbol{r}_{i+1}$ from this vector $\boldsymbol{r}$ in equation (3.35). If it is assumed that q_1 is now a unit test charge where forces throughout space may act upon it, the force per unit test charge may be defined as the *electric field*, or the force acting about a given charge q as

$$\vec{\boldsymbol{E}} = \frac{\vec{\boldsymbol{F}}}{q} = \frac{q(\boldsymbol{r}_1 - \boldsymbol{r}_2)}{4\pi\epsilon_0 |\boldsymbol{r}_1 - \boldsymbol{r}_2|^3} \tag{3.36}$$

The electric field $\boldsymbol{E}$ in equation (3.36) has units of volts/meter. If a charge q is defined in *Coulombs*, and the distances in meters, the constant k_e may be defined as

$$k_e = \frac{1}{4\pi\epsilon_0} \frac{N \cdot m^2}{C^2} \tag{3.37}$$

which is called *Coulombs constant*. Typically, the permittivity is separated out when computing problems that only involve fields rather than *force* vectors. In real world problems, the electric fields encountered are often sums of charges either on the surface of a conductor or within a certain medium. Thus, it is typical to use static values rather than fields to describe a given problem. If one tries to move a charge in space from its original position through a path and then back to the same point, it is found that no work has been done; or stated another way, if a charge is traversed through an electric field over some closed contour the total work done will be

$$W = \oint \vec{F} \cdot d\vec{l} = q \oint \vec{E} \cdot d\vec{l} = 0 \tag{3.38}$$

This implies that the following equation holds as well; that the integration of the static electric field around a closed contour will be zero:

$$\oint \vec{E} \cdot d\vec{l} = 0 \tag{3.39}$$

But this relation is true only for *static* fields, and it is defined as the *conservation of the electro-static* field. The notion is that if a charge is moved around a loop back to the same point, no work would have been done and so this is a form of *conservation of energy*. Nevertheless, in many cases it is of more interest to compute the electric field from one point to the next. Since the electrostatic field has units of volts/meter, it is apparent that by integrating from one point to another (or rather moving a charge or set of charges through a path), the electrostatic potential can be obtained between the two points:

$$V_a - V_b = -\int_a^b \vec{E} \cdot d\vec{l} \tag{3.40}$$

The *voltage*, of course, is a scalar quantity, as was noted earlier in the discussion of the *dot* product between two vectors. The negative sign in equation (3.40) comes, by definition, from Coulombs law, by which it is possible to define positive work when a charge is moved through an electric field toward the source (opposite in direction to the electric field vector). The voltage between two points is also called the *electromotive force*. This is because it is essentially moving electrons in the presence of a field and then summing the force over the *net* distance to accomplish this. Notice that for a static field, the path does not matter, only the actual distance that the charge (or charges) travels in parallel with the electric field vector. The differential form of equation (3.40) is

$$-\nabla V = \vec{E} \tag{3.41}$$

The electric field is a *vector* quantity, as defined in the previous section. This notation is used extensively for numerical computations [3].

The discussion up to this point has been on electro-static forces, charges, and electric fields. There is another term used to describe the density of fields within a region, and that is *flux*. The *electric flux density* vector $\boldsymbol{D}$ is a quantity that is independent of the permittivity,

$$\vec{D} = \epsilon_0 \vec{E} \tag{3.42}$$

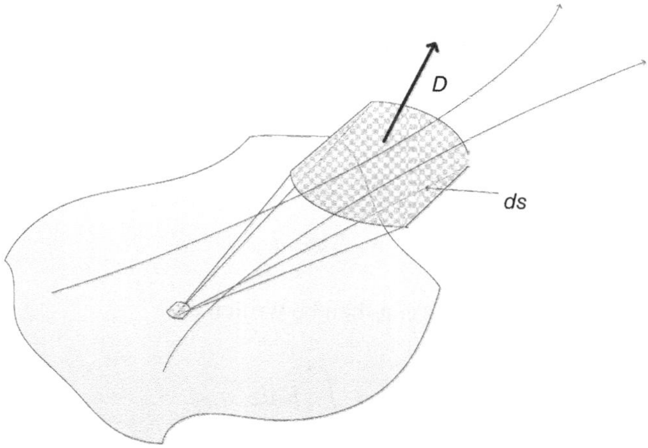

Figure 3.8 Electric flux density vector

The electric flux density is typically defined in terms of the total flux going into and out of a given surface. The electric flux density vector is related to the electric field through the dielectric constant of the material.

The total electric flux through a given region (as shown in Figure 3.8) may be determined by summing the electric flux density through a surface (i.e., adding the differential elements of that surface dS). For example, to find the total flux due to a set of charges through a surface S, the electric flux is integrated over that region:

$$\textit{Total electric flux} = \oiint \vec{D} \cdot d\vec{S} \tag{3.43}$$

where S is the total enclosed area of the surface. Equation (3.43) is very useful in that it is also equal to the total charge enclosed within the surface:

$$\oiint \vec{D} \cdot d\vec{S} = Q_{enclosed} \tag{3.44}$$

Using the divergence theorem obtains

$$\oiint \vec{D} \cdot d\vec{S} = \iiint \nabla \cdot \vec{D} dv = \mathbf{Q}_{enclosed} \tag{3.45}$$

or, in differential form,

$$\nabla \cdot \vec{D} = \rho \tag{3.46}$$

which is called Gauss's law. The usefulness of this relation may be illustrated by an example.

Example 3.3 Referring to the capacitor in Example 3.2, determine the equation for a parallel plate capacitor using the electrostatic potential equation. Assume that the lower plate is in the $z = 0$ plane and neglect any fringing field effects.

SOLUTION: First assume that the width and length of the capacitor are much greater than the plate separation d, or

$$w,\ l \gg d \tag{3.47}$$

The voltage across the capacitor can then be written as

$$\mathbf{V} = \int_0^d E_z dz \tag{3.48}$$

The electric field between the plates of the capacitor is determined using Gauss's law, by which the charge density is assumed to be constant throughout:

$$\nabla \cdot \vec{D} = \frac{dE_z}{dz}\epsilon_0 = \rho_v \tag{3.49}$$

Since $\epsilon_0 E_z = D_z$. After integrating both sides of equation (3.49) over z, the following relation obtains:

$$E_z = \frac{\rho_s}{\epsilon_0} \tag{3.50}$$

where the charge density ρ_s is now over the surface of the capacitor. After substituting for E_z in equation (3.48) the solution becomes

$$V = \int_0^d \frac{\rho_s}{\epsilon_0} dz = \frac{\rho_s d}{\epsilon_0} = \frac{Qd}{A\epsilon_0} \rightarrow Q = V\frac{\epsilon_0 A}{d} \rightarrow C = \frac{\epsilon_0 A}{d} \tag{3.51}$$

which is the equation for a parallel plate capacitor (in air). Note the substitution for the charge density in equation (3.51), which is simply the total charge divided by the area. This handy relation can be used over and over to determine the capacitance of various structures, or at least to approximate them.

Look back for a moment to the *force* equation and recall that the work done from moving a charge through an electric field is equal to the charge times the electrostatic potential and is equivalently zero if the charge is traversed through a closed path back to the same potential in the field. The total work (or energy stored in the electric field) [5] for n charges will then be

$$W_e = \frac{1}{2}\sum_{i=1}^{n} Q_i V_i \tag{3.52}$$

It follows that the total work stored in a capacitor can be found by substituting for the charge in equation (3.52). Now, suppose that there is a differential charge stored on a conductor (e.g., a capacitor) with a known voltage V across it, then by integration the work is

$$W = \int_0^Q V\,dq = \int_0^Q \frac{q}{C}dq = \frac{1}{2}\frac{Q^2}{C} = \frac{1}{2}CV^2 \tag{3.53}$$

This is the total energy stored in a capacitor. It follows from the divergence theorem that

$$W = \frac{1}{2}\iiint \nabla\cdot\vec{D}V\,dv = \frac{1}{2}QV \tag{3.54}$$

Then

$$W = \frac{1}{2}\oiint V\vec{D}\cdot ds - \frac{1}{2}\iiint \vec{D}\cdot(\nabla V)\,dv \tag{3.55}$$

after applying a vector identity (see the Appendix) to the middle integral in equation (3.54). This integral also decays to zero if the volume includes all space. Substituting in the last integral for the electric field $\boldsymbol{E}$ obtains the result arrived at below:

$$W = \frac{1}{2}\iiint \vec{D}\cdot\vec{E}dv = \frac{1}{2}\iiint \epsilon|\vec{E}|^2\,dv \tag{3.56}$$

This is a general equation that applies to any set of static fields in space and can be used to determine the static energy stored in a system where the charge accumulates, as it does in conductors.

Example 3.4 Find the capacitance in a coaxial structure of unit length l, with radii $\boldsymbol{r}_1$ and $\boldsymbol{r}_2$, using the energy relation and the divergence theorem. Again, neglect any fringing field effects in the computations.

SOLUTION: This is a classic problem in electromagnetics. The first step is to determine the electric field in radial coordinates while referring to Figure 3.9:

$$q = \oiint \vec{D}\cdot d\vec{S} = \int_0^l\int_0^{2\pi} \epsilon E_r r\,dr\,d\theta \tag{3.57}$$

After integration,

$$q = 2\pi\epsilon E_r\,rl \;\rightarrow\; E_r = \frac{q}{2\pi r\epsilon l} \tag{3.58}$$

Then, using the energy relation in equation (3.56) obtains

$$W = \frac{1}{2}\iiint \vec{D}\cdot\vec{E}dv = \frac{1}{2}\int_{r_1}^{r_2}\int_0^{2\pi}\int_0^l \epsilon\left(\frac{q}{2\pi r\epsilon l}\right)^2 r\,dr\,d\theta\,dl = \frac{q^2}{4\pi\epsilon l}\ln\left(\frac{r_2}{r_1}\right) \tag{3.59}$$

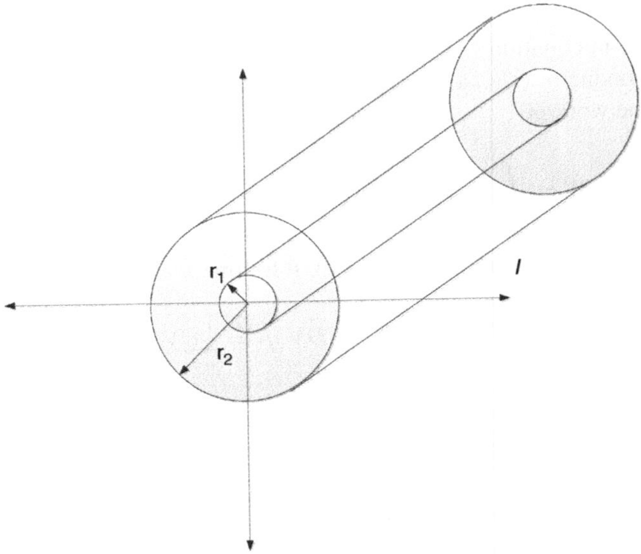

Figure 3.9 Coaxial capacitor problem

Last, equate it to the energy stored in a capacitor:

$$W = \frac{1}{2}CV^2 = \frac{(CV)^2}{4\pi\epsilon l}\ln\left(\frac{r2}{r1}\right) \rightarrow C = \frac{2\pi\epsilon l}{\ln\left(\frac{r_2}{r_1}\right)} \quad (3.60)$$

Which is the equation for a cylindrical capacitor.

3.2.2 Magneto-Statics

The study of magneto-statics is close to that of electro-statics except that the focus is on currents that generate fields instead of fields generated from charges. Ampere's law, which will be introduced shortly, is similar to Coulombs law except that it concerns *current-carrying elements* rather than charges.

Analogous to capacitance, in this case it is *inductance* that will be encountered as one of the fundamental circuit components. Recall that in chapter 2, energy may be stored in an inductor or used as a filter, or both, when it is necessary to convert power from one voltage to another—as in a buck regulator. Inductance is related to total flux, and the current is given by the relation

$$\psi = LI \quad \text{or} \quad L = \frac{\psi}{I} \quad (3.61)$$

The total flux ψ is generated from a field in a current-carrying conductor. The inductance for a particular structure will be determined later in this section.

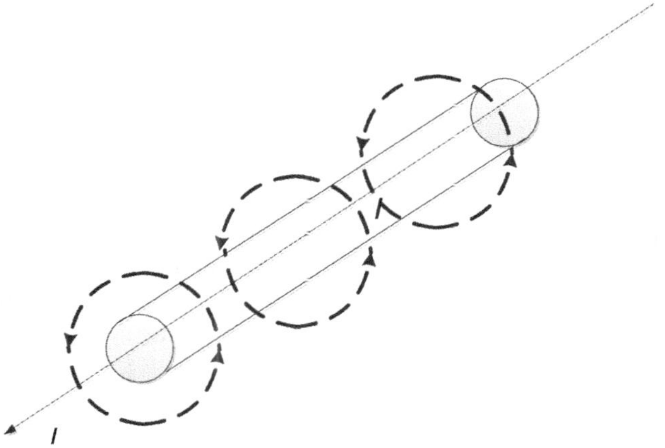

Figure 3.10 Field due to a current in a wire

The total flux, ψ is obtained from the magnetic flux density $\boldsymbol{B}$, which is a vector quantity. The magnetic flux density vector $\boldsymbol{B}$ is the density of flux per a given unit area in space (in this case). Consider Figure 3.10, where a wire has a current running through it generating a field in free space. As a current I is flowing in the wire shown in the figure, a *field* is generated about this wire. The quantity or measure of *field lines* or *flux lines*, which it comprises, is the total flux. The concept of flux was introduced when the *electric flux density* was discussed for static electric fields in an earlier section. The flux density for *magnetic* fields is $\boldsymbol{B}$ as defined above. The direction of that field is determined by the *right-hand rule*. If one wraps their right hand about a wire and places the thumb in the direction of the current, the direction of the field will be determined by how the fingers in the right-hand curl around that wire. The magnetic flux density vector is related to the *magnetic* field by the relation

$$\vec{\boldsymbol{B}} = \mu_0 \vec{\boldsymbol{H}} \tag{3.62}$$

where μ_0 is the *permeability* of the material. The SI units for the *magnetic flux density* $\boldsymbol{B}$ is the *Tesla*. In general, permeability is a *tensor* product for most magnetic materials and has a relationship that is directional with respect to the magnetic field's direction and strength.

The permeability constant μ_0 in free space has the value

$$\mu_0 = 4\pi \times 10^{-7} \frac{\text{H}}{\text{m}} \tag{3.63}$$

where H stands for Henrys. For many passive structures in a printed circuit board, there is no magnetic material, and thus for most of the problems, such as those that will be encountered here, knowing the value of this constant is sufficient. Ampere's law may now be introduced so that we can relate the magnetic field to the current.

The total current in a conductor is found by integrating the total magnetic field around a closed loop, expressed in integral form as

$$I = \oint \vec{H} \cdot dl \tag{3.64}$$

The integral in equation (3.64) around the closed loop is a *line integral*. This result was derived from a set of experiments and is very useful in determining magnetic parameters for power integrity problems—for example, the inductance for a type of conductor geometry. The total flux is thus found by integrating the magnetic flux density over an area bounded between two conductors as in Figure 3.9. The total flux is

$$\psi = \oiint \vec{B} \cdot dA \tag{3.65}$$

This is the static form of *Faraday's law of induction*. Faraday's law states that the magnetic flux through a wire loop is proportional to the number of magnetic flux lines that pass through that loop. The dA in equation (3.65) is the infinitesimal differential area in that loop. Faraday's law applies to time-changing fields and regions (moving loops) as well as to static fields and regions. The use of the previous two formula's are now illustrated by an example.

Example 3.5 Find the inductance in the coaxial structure in Figure 3.9 using Ampere's and Faraday's laws—again, ignore any fringing field effects.

SOLUTION: First determine the magnetic field from Ampere's law:

$$I = \oint \vec{H} \cdot dl = \oint_0^{2\pi} H_\phi r d\phi \tag{3.66}$$

Integrating obtains

$$I = 2\pi r H_\phi \;\rightarrow\; H_\phi = \frac{1}{2\pi r} \tag{3.67}$$

Now substitute for the magnetic flux density in Faraday's equation, and ignore fringing fields:

$$\psi = \oiint \vec{B} \cdot dA = \int_{r_1}^{r_2} \int_0^l \mu_0 H_\phi dr dl = \int_{r_1}^{r_2} \int_0^l \mu_0 \frac{I}{2\pi r} dr dl = \frac{\mu_0 I l}{2\pi} \ln \left(\frac{r_2}{r_1}\right) \tag{3.68}$$

Since

$$\psi = LI \tag{3.69}$$

$$L = \frac{\psi}{I} = \frac{\mu_0 l}{2\pi} \ln \left(\frac{r_2}{r_1}\right) \; H \tag{3.70}$$

Note, the same expression is used to find the *inductance per unit length* (which is common for transmission line extractions), but without variable l, which is omitted.

For the magnetic flux density vector $\boldsymbol{B}$, the question often arises: What forces exist from the fields generated from the currents in a conductor? It is understood that the product of a differential element $\boldsymbol{dl}$ in space, with a current I and the magnetic flux density $\boldsymbol{B}$, exerts a differential force $\boldsymbol{df}$. Or in vector notation,

$$\boldsymbol{df} = I\boldsymbol{dl} \times \boldsymbol{B} \tag{3.71}$$

The right-hand side forms a vector product where the direction of the vector $\boldsymbol{df}$ is orthogonal to plane that $\boldsymbol{dl}$ and $\boldsymbol{B}$ resides in and is determined by the *right-hand rule*. This relation is analogous to that determined for the force exerted from a charge in space. Thus, it can be seen that a current will also exert a force determined by its direction flowing through a conductor and the vector direction of the magnetic flux vector $\boldsymbol{B}$. This is illustrated in Figure 3.11.

If the differential current is moving through a magnetic field, as a charge through an electric field as expected, *work* is created in the process. As in the discussion on electrostatics, energy is also stored in the magnetic field. Recall from the earlier discussion on energy stored in an inductor for power conversion, as in a buck converter, that energy can be stored in *any* inductor. The following energy relation from Chapter 2 applies to any inductor through which a current flows:

$$W_m = \frac{1}{2}LI^2 = \frac{\psi}{2L} \tag{3.72}$$

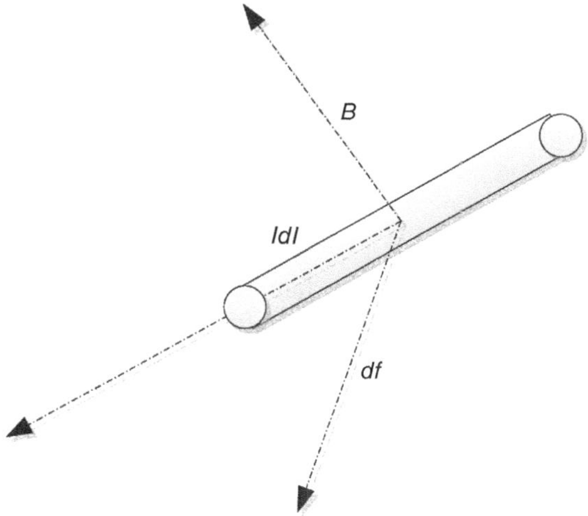

Figure 3.11 Transmission of differential current and corresponding forces in space

Analogous to the energy stored in an electric field, for magnetic fields,

$$W = \frac{1}{2}\iiint \vec{B}\cdot\vec{H}dv = \frac{1}{2}\iiint \mu|\vec{H}|^2\, dv \tag{3.73}$$

Equation (3.73) is a well-known expression for energy stored in a magnetic field, and follows directly from the circuit relation for the flux. At the end of this chapter (and in the appendix), these formulas will be utilized for some of the more common inductance geometries. These will help elucidate some of the more common derivations used for trace and power plane geometries of interconnections.

3.2.3 Conduction and Resistance

The third topic on static fields involves fields and currents that are *inside* of the conductor. When a current flows through a conductor, electrons collide, and this causes them to propagate in a direction (as a group) down the conductor itself. This concept is called current. Current is the amount of charge that passes through a region of a conductor in some unit of time, or

$$I = \frac{dq}{dt} \tag{3.74}$$

which is commonly understood from the study of electric circuits. The current density through a region of a conductor is related to the electric field:

$$\mathbf{J} = \sigma\mathbf{E} \tag{3.75}$$

where $\mathbf{J}$ is the *current density vector* and has units of amps/m^2. The term σ is called the *conductivity* of the material. For copper, the conductivity is

$$\sigma \cong 5.7 \times 10^7\ \frac{\text{S}}{\text{m}} \tag{3.76}$$

where the dimension S is called the *Sieman*. The conductivity is the inverse of resistivity

$$\sigma = \frac{1}{\rho} \tag{3.77}$$

which is in Ω/m and is usually the more familiar term when it is desired to compute the resistance of a structure. The value given for the conductivity (or the inverse of resistivity) is determined at 20°C, or roughly at room temperature. Different metals have different conductivities, and the constants can be easily found in the literature or on the Internet.

The total current is determined by integrating $\boldsymbol{J}$ over the area where the current passes through:

$$\oiint \boldsymbol{J}\cdot d\boldsymbol{S} = \mathbf{I} \tag{3.78}$$

where dS is the differential surface of the area. The conductivity of a copper structure is of special interest to the power integrity engineer because copper's conductivity is finite and usually results in power loss in the system. From the previous discussions on resistors, the power loss in a conductor is found as

$$P_{loss} = I^2R = VI \tag{3.79}$$

where the expression is in *watts*. The power loss per unit volume can be found by taking the dot product of the electric field $\boldsymbol{E}$ and the current density vector $\boldsymbol{J}$:

$$\frac{dP}{dv} = \boldsymbol{E} \cdot \boldsymbol{J} \ \frac{\mathrm{W}}{\mathrm{m}^3} \tag{3.80}$$

The total power then is found by integrating that expression over the volume V:

$$P = \iiint \boldsymbol{E} \cdot \boldsymbol{J}\, dv \tag{3.81}$$

Equation (3.81) is known as *Joule's* law. An illustration of the use of this equation follows.

Example 3.6 Find the resistance of the circular copper wire in Figure 3.12. Assume that the current is uniform across the conductor and that the length of the wire is l.

SOLUTION: Because the current is uniform, the current and current density are constants:

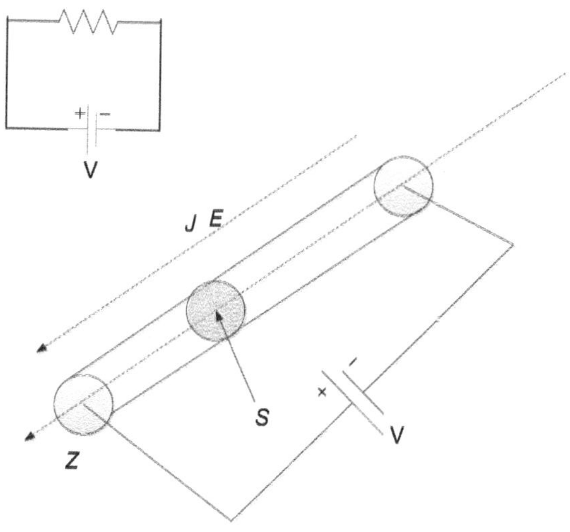

Figure 3.12 Copper wire in space

$$I = \oiint J_z \, dS = J_z \pi r_1^2 \tag{3.82}$$

Note the current in the z direction in the figure as this is the electric field. The electric field is found from the current and is also a constant:

$$V = -\int E_z \, dl = E_z l \tag{3.83}$$

The voltage and current is equivalent to the power lost in the conductor.

$$P = VI = J_z \pi r_1^2 \cdot E_z l = \frac{(J_z)^2}{\sigma} \pi r_1^2 l = I^2 R \tag{3.84}$$

Substituting for the current I obtains the following relation:

$$I^2 R = \frac{I^2}{\pi r_1^2 \sigma} l \rightarrow R = \frac{l}{\pi r_1^2 \sigma} \tag{3.85}$$

which is the total resistance for the circular wire.

3.3 MAXWELL'S EQUATIONS

Maxwell's equations are now discussed for completeness. For over a century, Maxwell's equations have formed the basis of the field and circuit relations used in electrical engineering disciplines. When some problem is encountered that is related to electricity, it is rare that Maxwell's equations do not show up in one form or another. Though it is not clear if James Clerk Maxwell intended, or even foresaw, to what extent that these equations would useful, today without the clarity that they express mathematically and symbolically, many great advancements—including the computer—would not have occurred as rapidly. For the power integrity engineer, these are fundamental equations as both circuit and field behavior are based upon them. Static field concepts have already been introduced as derived from Maxwell's equations, and for most power integrity problems, the static approximations will more than suffice. Thus, only the general form of Maxwell's equations will be illustrated here; it is up to the interested reader to investigate these equations further in the references listed at the end of the chapter.

Maxwell's equations are applicable to both time and spatial variations, and taken together, they create the fundamental basis that allows engineers to manipulate the math and physics behind the electrical behavior of virtually an infinite number of electrical structures that exist.[1] Maxwell's equations in *differential* form, and the

[1] Maxwell's contribution was to add the middle term in Ampere's law. Basically, Maxwell postulated that not only will a magnetic field be generated by an electric current but also by a time-changing electric field.

MKS system, are shown in equations (3.86) through (3.89) below:

$$\nabla \times \vec{\mathbf{H}} = \frac{\partial \vec{\mathbf{D}}}{\partial t} + \vec{\mathbf{J}} \tag{3.86}$$

$$\nabla \times \vec{\mathbf{E}} = -\frac{\partial \vec{\mathbf{B}}}{\partial t} \tag{3.87}$$

$$\nabla \cdot \vec{\mathbf{B}} = 0 \tag{3.88}$$

$$\nabla \times \vec{\mathbf{D}} = \rho \tag{3.89}$$

Again, note that these have already been seen in simpler form. Equation (3.86) gives *Ampere's law*, which states that the curl of the magnetic field **H** is equal to the time-rate of change of the electric flux density **D** and the current density **J**. It basically describes the relationship between the magnetic and electric fields. The dimensions are in A/m^2. Equation (3.87) is *Faraday's law of induction*. Through a variety of clever experiments, Faraday observed that a magnetic field induces an electric field, or an EMF (electro-motive force), from one circuit to another. In one such experiment, Faraday placed an iron bar into and out of a coil and watched a galvanometer needle move, thus, showing that through induction that an electric field can be formed. The dimensions for Faraday's law are in volts/m^2. Equation (3.88) is *Gauss's law for magnetism*. Gauss basically states that there are no magnetic monopoles or magnetic-free charges. There are instead only *magnetic dipoles* that are represented by loops of current rather than by free charges as was shown in earlier sections where charges that generate electric fields were discussed. Equation (3.89) is *Gauss's law* that relates the surface in which an electric flux passes through and the charges it encloses. This is similar to the divergence theorem. The dimensions for this equation are C/m^3, as seen in the charge density on the right-hand side of the equation.

In 1865, Maxwell published his document on electromagnetism that essentially unified the equations, based on the theory that electromagnetic fields travel through space and at the speed of light. This discovery is historically a major advancement in physics.

3.3.1 The Wave Equation

Together, the four Maxwell equations describe succinctly *macro-scopic* physics, which studies the manipulation of electromagnetic fields. As Maxwell observed, the solutions to these equations take the form of electromagnetic waves propagating in various media. Most waves are represented as *time-harmonic* waves; that is, the fields and waves change with frequency and time as well as spatially as shown for the electric field below:

$$\mathscr{E}(x, y, z, t) = Re\{\vec{\mathbf{E}}(x, y, z)e^{j\omega t}\} \tag{3.90}$$

where the script symbol for the electric wave on the left-hand side represents a *time-harmonic signal*. The real part of the electromagnetic field is taken here and the time-dependency separated out to represent that portion in a time-harmonic fashion. Not only the electric field but *all* the vectors from Maxwell's equations can be represented in this way [6]. Most signals that are generated from power sources are TEM (*transverse electromagnetic*), or quasi-TEM, due the fact they are generated on a printed circuit board and are of relatively low frequency. As in Faraday's law, equation (3.87), the curl of both sides can be taken by assuming that the fields are time-harmonic:

$$\nabla \times (\nabla \times \vec{\mathbf{E}}) = \nabla \times \left(-\frac{\partial \vec{\mathbf{B}}}{\partial t}\right) \rightarrow \tag{3.91}$$

$$\nabla \times (\nabla \times \vec{\mathbf{E}}) = -\mu \frac{\partial}{\partial t}(\nabla \times \vec{\mathbf{H}}) \tag{3.92}$$

where the permeability has been taken out and the derivative can now be found with respect to time, due to the time-harmonic assumption, without loss in generality. Using one of the vector identities from the Appendix on the left-hand side of the equation and then replacing the curl relation on the right, obtains

$$\nabla^2 \mathbf{E} - \nabla(\nabla \cdot \mathbf{E}) = \mu\epsilon \frac{\partial^2 \mathbf{E}}{\partial t^2} \tag{3.93}$$

where the vector sign (again) has been omitted for simplicity and a *lossless* medium has been assumed for the moment. If there are no free charges in this space, the middle term goes to zero—which is known from Gauss's law since $\rho = 0$. Thus, by the fact that the solution to the equation is time-harmonic or *sinusoidal* in nature, the final result becomes

$$\nabla^2 \mathbf{E} - \omega^2 \mu\epsilon \mathbf{E} = \nabla^2 \mathbf{E} - \mathrm{k}^2 \mathbf{E} = 0 \tag{3.94}$$

which is the *wave equation* for a lossless medium (for an electric field, a similar one may be derived for the magnetic field). The variable k on the right side of equation (3.94) is called the *wave number* (or sometimes the *phase constant*), and it describes the magnitude of the *wave vector*. The wave number has units of 1/m. The wave number may be represented in terms of angular frequency and velocity:

$$k = \frac{\omega}{v} = \omega\sqrt{\mu\epsilon} \rightarrow v = \frac{1}{\sqrt{\mu\epsilon}} \tag{3.95}$$

For free space, the velocity of the wave is c, or the speed of light:

$$v_{fs} = c = \frac{1}{\sqrt{\mu\epsilon}} \cong 3.00 \times 10^8 \ \frac{\mathrm{m}}{\mathrm{s}} \tag{3.96}$$

where v_{fs} is the velocity of a wave in free space. The wavelength of the signal can be computed from the wave number and phase velocity:

$$\lambda = \frac{2\pi}{k} = \frac{2\pi}{\omega\sqrt{\mu\epsilon}} = \frac{2\pi}{2\pi f\sqrt{\mu\epsilon}} = \frac{v}{f} \tag{3.97}$$

From a power integrity perspective, the waves within the printed circuit board are usually generated from dynamic load changes and/or the power converter switching itself. As will be seen in later chapters, the frequency band of interest for power integrity applications will typically be less than 1 GHz. Thus, static field concepts can be used to approximate the actual structure in most cases. However, for some problems, more accurate formulas are needed to get the correct results. This is particularly true when operating in media where losses are significant.

Example 3.7 Compute the wavelength of a 100 MHz signal if the signal is TEM and propagates in an FR-4 board with permittivty constant of 3.6.

SOLUTION: Begin by assuming that the signal is propagating in a uniform dielectric. Because the current is uniform, the current and current density are constants. Using equation (3.97), we obtain

$$\lambda = \frac{v}{f} = \frac{1}{100e^6\sqrt{\mu\epsilon_r\epsilon}} = \frac{300e^6}{100e^6\sqrt{3.6}} = \frac{3}{1.89} = 1.58 \text{ m} \tag{3.98}$$

The result is in meters, as expected.

3.3.2 Lossless and Lossy Media

To this point, the discussion has been primarily about lossless media. That is, when a signal is sent through a medium or power on a structure, it exhibits little or no loss. In practice, however, real conductors are resistive and power is dissipated. If a signal or wave is transmitted into a lossless medium, it will not dissipate any *real* power. However, if the medium is *lossy*, then power loss will be observed on that signal. To see how this occurs, a revisit to Ampere's law on sinusoidally varying waves is appropriate. If there is no conduction losses, than Ampere's law becomes

$$\nabla \times \mathbf{H} = j\omega\epsilon\mathbf{E} + \sigma\mathbf{E} = j\omega\epsilon\mathbf{E} \;\rightarrow\; \sigma = 0 \tag{3.99}$$

For lossy media, the solution is now

$$\nabla \times \mathbf{H} = (j\omega\epsilon + \sigma)\mathbf{E} = j\omega\epsilon\left(1 - j\frac{\sigma}{\omega\epsilon}\right)\mathbf{E} \tag{3.100}$$

The second term in equation (3.100) is called the *loss tangent*:

$$\tan\phi = \frac{\sigma}{\omega\epsilon} \tag{3.101}$$

The loss tangent for many materials can be found in tables on the Internet or in many texts. For a very good conductor, it is found that

$$\frac{\sigma}{\omega\epsilon} \gg 1 \tag{3.102}$$

Returning to the wave number, for a moment, in equation (3.95), if attenuation is added to this constant, then the new metric can be broken into a *real* and a *complex* component:

$$\gamma = \alpha + j\beta \tag{3.103}$$

where α is called the *attenuation constant* and β is the phase constant (same as k) and gamma, γ, is called the propagation constant. If the attenuation constant is solved for, and it is assumed that the wave is traveling in a *good* conductor, the result is

$$\alpha \cong \sqrt{\frac{\omega\mu\sigma}{2}} \tag{3.104}$$

The inverse of this is called the *skin-depth*:

$$\delta = \sqrt{\frac{2}{\omega\mu\sigma}} = \frac{1}{\sqrt{\pi\mu f\sigma}} \tag{3.105}$$

This is the *thickness* of a conductor, a signal will propagate into the thickness at a given frequency to maintain its magnitude proportional to e^{-1}, or 36.8% of its maginitude. Skin-depth was discussed briefly in Chapter 2 to describe frequency dependent losses in inductor designs or interconnects. More often it is used to determine the penetration of a current waveform into conductor (e.g., copper) at a given frequency. When determining the losses in a copper plane as an AC waveform is propagating into it, it is sometimes best to do a check on the skin-depth to ensure that the losses are manageable.

Example 3.8 Determine the *skin depth* for the signal in Example 3.6 at 20°C in a copper plate. If the plate is 1 mm thick, will there be additional loss; that is, will the current penetrate in the copper wire in less than the thickness given?

SOLUTION: All of the values for the variables are shown below. The temperature affects only the conductivity of copper so the value below is correct,

$$\begin{aligned} \sigma &\cong 5.7 \times 10^7 \ \frac{\mathrm{S}}{\mathrm{m}} \\ f &= 100 \times 10^6 \\ \mu_0 &= 4\pi \times 10^{-7} \ \frac{\mathrm{H}}{\mathrm{m}} \end{aligned} \tag{3.106}$$

Thus,

$$\delta = \frac{1}{\sqrt{\pi \mu f \sigma}} = \frac{1}{\sqrt{\pi (5.7 \times 10^7)(100 \times 10^6)(4\pi \times 10^{-7})}} = 2.11\ \mu\text{m} \quad (3.107)$$

If the current penetrates from both sides, then the total current density will be 4.22 μm deep. This is 250 times smaller than the copper plate! There will be significant loss due to the signal because the current will reside (mainly) within the skin-depth of the metal.

3.4 USEFUL AND SIMPLE CIRCUIT EXTRACTIONS

It has been shown to this point that some of the basic concepts may help in solving some basic power integrity first principal circuit extractions. In this last section, an introduction to a few more common extractions is given to aid the reader. The focus here will be on a few inductance extractions and one resistance extraction. Since capacitance is usually small for many of the problems encountered in power integrity, this will be left for chapter four and in some exercises for the reader to follow. The intention is for the PI engineer to use these formulae and concepts as problems present themselves. These are fairly common closed-form results (as has been shown in this chapter) and may be found in a number of texts and even on the internet if one is willing to look for them. They are provided here with some clarity for the reader.

3.4.1 "Power Plane" Inductance

One of the more common circuits encountered is the inductance of a power plane [8]. The *first principles* approximation is straightforward, and the result can be used in many common problems in power integrity. Even for estimates of geometries that are somewhat complex, this equation can be applied. Refer to Figure 3.13 where two parallel plates are separated in space. Take note of the following parameters:

$$s \ll w,\ l \quad (3.108)$$

where s is the separation of the plates, w is the width of each plate, and l is the length. A current i flows in the z direction, shown on the top plate, with the return current in the opposite direction flowing on the bottom. The current is assumed to be uniform throughout the plate, and fringing fields are once again neglected. The goal is to compute the inductance of the plates given the dimensions in the cartesian coordinate system. The first goal is to determine the magnetic field in the z direction:

$$I = \oint \vec{H} \cdot dl = \int_0^w H_z \cdot dx + \int_w^0 H_z \cdot dx \quad (3.109)$$

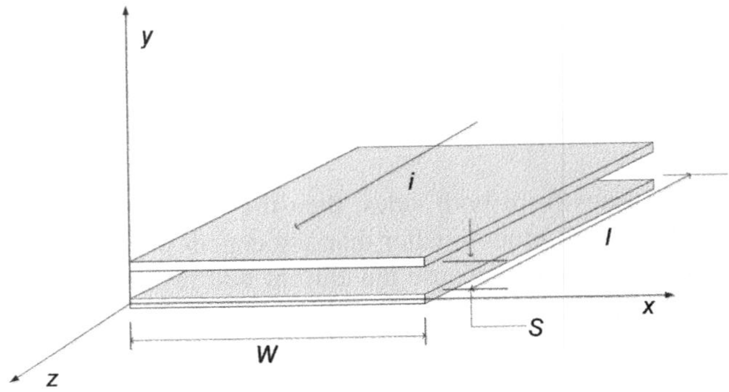

Figure 3.13 Two parallel plates in space for inductance computation

where the integration around the edges (remember, the integration loop is around the conductor that contains the current) has been neglected. Since current flows in equal and opposite directions, the temptation is to integrate around the one conductor and assume that there is an H_z directed field on the top side that can also be added to the total enclosed current. However, due to the current in the $y = 0$ plane, it is found that this magnetic field can be neglected—since the magnetic field above the top conductor is essentially zero by definition (e.g., the width and length of the plate are large relative to the separation). Using these approximations, it is determined that

$$I \cong \int_0^w H_z \cdot dx \;\rightarrow\; H_z = \frac{I}{w} \tag{3.110}$$

Now, substitute for H_z in Faraday's equation, as was done in Example 3.5. The integration is through the cross-sectional area between the plates to determine the total flux through that region:

$$\psi = \oiint \vec{B} \cdot dA = \int_0^l \int_0^s \mu_0 H_z dz dy = \int_0^l \int_0^s \mu_0 \frac{I}{w} dz dy = \frac{\mu_0 I l s}{w} = LI$$
$$\rightarrow\; L = \frac{\mu_0 l s}{w} \tag{3.111}$$

This result can be used for many simple approximations, and it is an invaluable addition to the power integrity engineer's tool set. As usual, it is important to check the assumptions first, particularly the frequency of operation. If there is *well developed* skin-depth, this relation will not hold and the inductance due to the skin-depth (or *internal inductance* in this case) must be added [8].

3.4.2 Inductance of Two Circular Wires in Space

An examination of two common computations for two parallel wires in space is discussed where each wire is assumed to be long relative to the radii and the

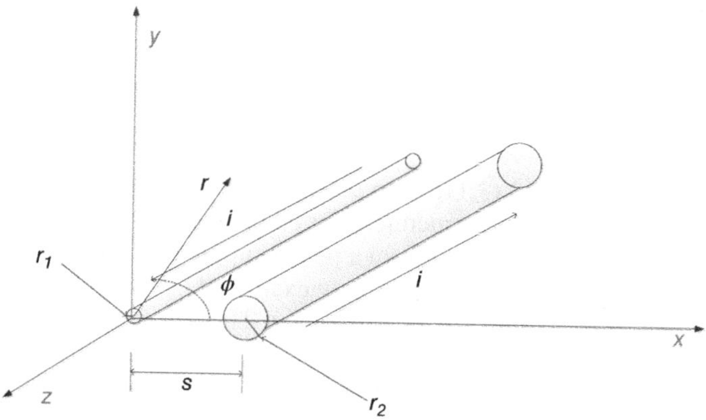

Figure 3.14 Two parallel wires of different radii in space

distance of separation, while, additionally, the separation is greater than either of the two radii [16]. The first assumption is given so that the magnetic fields are, for all intent and purposes, orthogonal to the direction of the current flow. This allows the use of static approximations (again, it is assumed the current is also static). Refer to Figure 3.14. Note that the wires have different radii. This is the general of the two cases that will be solved.

Because each wire carries a current i, the objective is to find the total contribution of flux from that current. Again, the assumptions are given first:

$$s,\ r_1,\ r_2 \ll l \quad \text{and} \quad r_1,\ r_2 < s \tag{3.112}$$

It is assumed that both wires can be approximated as perfectly conducting; that is, the fields will not penetrate the wires (e.g., no skin-depth). For very good conductors, this is a reasonable assumption. Additionally, it is assumed, in the circuit relation, that each wire is connected through a closed circuit on each end, resulting in the same current through each wire. As in Example 3.5, and using cylindrical coordinates, computing the total flux contribution from each wire yields

$$\psi = \oiint \vec{B} \cdot d\vec{A} = \int_{r_1}^{s-r_1} \int_0^l \mu_0 H_\phi dr dl + \int_{r_2}^{s-r_2} \int_0^l \mu_0 H_\phi dr dl \tag{3.113}$$

Notice that the integration from the second wire starts at r_2. This is because the direction of integration is from that edge of the second conductor. Integrating equation (3.113) yields

$$\frac{\psi}{I} = L = \frac{\mu_0 l}{\pi} \ln\left[\frac{(s-r_1)(s-r_2)}{r_1 r_2}\right] \tag{3.114}$$

The special case where the radii are equivalent yields the second equation:

$$L = \frac{\mu_0 l}{\pi} \ln\left(\frac{s-r}{r}\right) \tag{3.115}$$

Both equations (3.114) and (3.115) are useful for approximating the inductance between wires, such as in cable structures. For power integrity problems, this could mean a pair of power cables routed from a power supply to a printed circuit board.

A variant of the previous structure is now examined. As shown in Figure 3.15, a small wire sits above a large or infinite plane. The spacing between the plane and the wire (to the center) is h. The goal is to determine the inductance of the wire over this plane. If the current is uniform around the wire (as in the previous assumptions in this section), then the plane may be replaced by an *image* wire [5] at an equal distance h from the bottom of the plane. It is also assumed that the plane is negligibly thin so that the magnetic field can be determined in a similar fashion to the previous equations as

$$\oint_0^{2\pi} H_\phi r d\phi \rightarrow H_\phi = \frac{I}{2\pi r} \tag{3.116}$$

which is the same result arrived at previously. The objective is to now compute the total flux.

$$\psi = \oiint \vec{B} \cdot d\vec{A} = \int_{r_1}^{h-r_1} \int_0^l \mu_0 H_\phi dr dl \tag{3.117}$$

Note that the integration is only through *half* of the area. The resultant inductance is now

$$L = \frac{\mu_0 l}{2\pi} \ln\left(\frac{h-r_1}{r_1}\right) \tag{3.118}$$

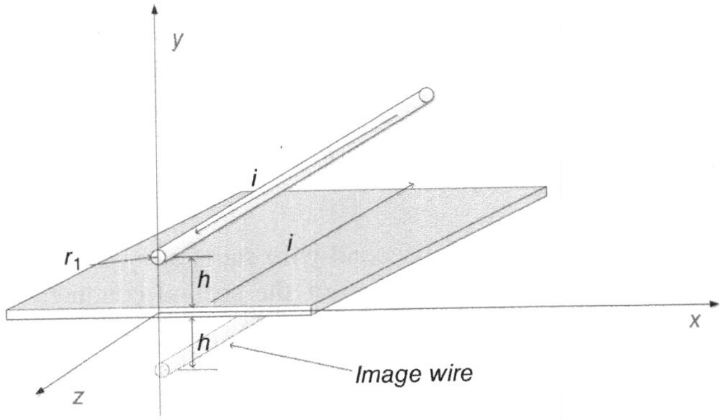

Figure 3.15 Wire over a large or infinite plane

It should be remembered that for all problems typically encountered in power integrity (with the possible exception of some canonical problems), the results are *approximations* to the actual. So the use of a good field solver is desirable where the physical construction is complex, since then any closed form estimation would be too inaccurate to use as an estimate.

3.4.3 Resistance between Two Vias in a Power Plane

The next problem is a simple resistance computation. Suppose that the copper sheet is large in the x–z direction and that the separation between the vias is large compared with the radius of each via. Refer to Figure 3.16. Note that each via has a diameter r and that the current flows from one via to the next. Faraday's law can be used to compute the voltage from the via to some point r in the conductor:

$$V = \int E_{\phi}\, rdr = \int \frac{J_{\phi}}{\sigma}\, rdr = \frac{J_{\phi}}{\sigma} 2\pi r \rightarrow J_{\phi} = \frac{V\sigma}{2\pi r} \tag{3.119}$$

where cylindrical coordinates are assumed once again. The total current can be found by integrating over the area from one via to the next:

$$I = \oiint J_{\phi} \cdot dS = \int_{r}^{s-r} \int_{0}^{t} \frac{V\sigma}{2\pi r}\, drdt = \frac{V\sigma t}{2\pi} \ln\left(\frac{s-r}{r}\right) \tag{3.120}$$

Now, the resistance between the vias is simply the voltage divided by the current:

$$R = \frac{V}{I} = \frac{2\pi}{\sigma t \ln\left(\frac{s-r}{r}\right)} \tag{3.121}$$

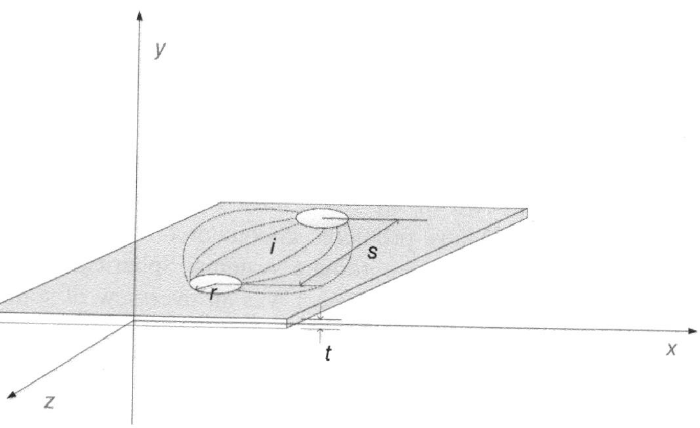

Figure 3.16 Two vias in a copper plane

Note, first, that in equation (3.121) the vias were assumed to be of equal size—this is a special case of the more general result where the via sizes are *unequal* in size. Second, the area of integration in equation (3.120) is the thickness times the differential radius. This only holds if the separation distance is somewhat large relative to the radius—typically, this number should be greater than 5 [16]. This result is handy for many problems where the objective is to determine the current into a via from a given source on a plane. Often engineers tackle problems where the current density exceeds the specifications of the via and issues such as *electro migration* come up. Some of these issues will be discussed in later chapters.

3.4.4 Notes on Applicability of Formulas

The formulas covered in this chapter—as well as those in the Appendix—are closed-form estimations for specific problems and thus have assumptions that must be heeded. For example, equation (3.115) is valid only if the separation between the wires is greater than five times the radius of the wire itself. Though, in most cases, when the formula is being used as an estimation, it is possible to violate these limits *slightly* to determine a general value. Nonetheless, the reader should be careful when applying any closed-form formula without having a clear understanding of its limitations. Documenting the limits and discussing every possible boundary error is beyond the scope of this book. Thus, the reader should consult the references at the end of the chapter if there is any doubt as to how to apply these concepts correctly. In the upcoming chapters, a number of these formulas will be reused and some discussion on their limits will be given.

3.5 SUMMARY

In this chapter, a number of basic concepts for electromagnetic structures were reviewed as a basis for the upcoming power integrity analyses. This introduction was limited because only discussed concepts are applicable overall to the subject at hand. The reader is reminded to consult the references at the end of this chapter for a more thorough study.

The chapter began with a review of vectors as used in the vector calculus. The discussion then moved to two of the three most common coordinate systems: *cartesian* and *cylindrical*. Many problems encountered in the study of power integrity deal with just these two coordinate systems. Spherical coordinates are discussed in the Appendix. The next subject was an overview of some common vector operations and vector calculus formulas, including the *divergence theorem* and *Stoke's theorem*, to aid the reader in understanding the upcoming formulas. The main concepts of this chapter deal with static fields, and especially the electrostatic and magneto-static fundamentals relevant to extracting circuit elements using *first principles* analyses. The number of important relations discussed included *Ampere's law* for static fields, *Faraday's law* and *Gauss's law*. In the electrostatic section, the discussion focused on the *charge* along the electric field and how work is done by an electric field, which led to the introduction of the electric flux vector ***D***. Additionally, the concept of *capacitance* was discussed in some detail, and a few examples were given on how to compute this important metric. In the m*agneto-static* section, the analyses of flux in magnetic fields helped introduce the magnetic density vector ***B***. The concept of inductance was introduced and some examples showed how to compute it using *first principles* for a few basic problems. Finally, the subject of conduction, current density, and resistance was addressed and current density vector ***J*** was defined.

In the last section, also Maxwell's equations and the wave equation were introduced along with a short overview on *skin-depth* and the *wave number*. The chapter concluded with some remarks on commonly used circuit extractions whose applicabilities will be explored in later chapters. A table of some common geometries for inductance is provided in the Appendix.

REFERENCES

1. Chew, C. C., et al. Fast solution methods in electromagnetics. *IEEE Trans. Antennas Propag.*, vol. 45, no. 3, 1997.
2. Mittra, R. Whiter computational electromagnetics? A practiioners look at the crystal ball. IEEE Computation in Electromagnetics Conference, 2008.
3. Zhu, Y., and Cangellaris, A. C. *Multigrid Finite Element Methods for Electromangetic Field Solving*. Wiley, 2006.
4. Sadiku, N. O. *Numerical Techniques in Electromagnetics*. CRC, 1992.
5. Paul, C. R., and Nasar, S. A. *Introduction to Electromagnetic Fields, 2nd ed.* Wiley, 1987.
6. Balanis, C. A. *Advanced Engineering Electromagnetics*. Wiley, 1989.
7. Marshall, S. V., DuBroff, R. E., and Skitek, G. G. *Electromagnetic Concepts and Applications*, 4th ed. Prentice Hall, 1996.
8. Hayt, W. H. *Engineering Electromagnetics*, 4th ed. McGraw-Hill, 1981.
9. Ramo, S., Whinnery, J. R., and Van Duzer, T. *Fields and Waves in Communication Electronics*, 3rd ed. Wiley, 1994.
10. Goddard, K. F., Roy, A. A., and Sykulski, J. K. Inductance and resistance calculations for isolated conductors. *Science, Measurement and Technology, IEE Proc.*, Vol. 152, no. 1, 2005.
11. Kim, J., et al. Inductance calculations for plane-pair area fills with vias in a power distribution network using a cavity model and partial inductances. *MTT, IEEE Trans.*, vol. 59, no. 8, 2011.
12. Ruehli, A. E. Inductance calculations in complex integrated circuit environment. *IBM J. Res. Dev.*, vol. 16, no. 5, 1972.
13. Grover, F. W. *Inductance Calculations*. Dover, 1952.

14. Hernandez-Sosa, G., and Sanchez, A. Analytical calculation of the equivilent Inductance for signal vias in parallel planes with arbitrary P/G via distribution, in complex integrated circuit environment. ICCDCS, 2012.
15. Zhou, F., Ruehli, A. E., and Fan, J. Efficient Mid-frequency Plane Inductance Calculation. POWERCON, 2010.
16. Paul, C. R. *Introduction to Electromagnetic Compatibility*. Wiley, 1992.
17. Leferink, F. B. J. Inductance calculations: Methods and equations. *IEEE Int. Symp. Electromagnetic Compatibility Symposium Record*, 1995. pp. 16–22.
18. Leferink, F. B. J., Van Doom, M. J. C. M. Inductance of printed circuit board ground planes. *IEEE Int. Symp. Electromagnetic Compatibility Symposium Record*, pp. 327–329, 1993.
19. Leferink, F. B. J. Inductance calculations: Experimental investigations. *IEEE Int. Symp. Electromagnetic Compatibility Symposium Record*, pp. 235–240, 1996.
20. Hoer, C., and Love, C. Exact inductance equations for rectangular conductors with applications to more complicated geometries, *J. Res. National Bureau of Standards-C Engineering and Instrumentation*, vol. 69C, no. 2, 1965.

PROBLEMS

3.1 If vector **A** is subtracted from vector **B** in Example 3.1, and the y element is multiplied by −1.4, what are the magnitude and angle of the new vector **C**?

3.2 Determine, using Stoke's theorem, the surface integral if vector $\boldsymbol{F}$ is

$$\vec{\boldsymbol{F}} = 3y^2\boldsymbol{u}_x + 7xz\boldsymbol{u}_y + yx^2\boldsymbol{u}_z \tag{3.122}$$

3.3 Find the electric field $\boldsymbol{E}$ for the capacitor in Example 3.2. What is the total force?

3.4 If the electric field is given in the equation and positions for V_a and V_b as (2, 1, 4) and (−3, 7, 9), respectively (in mm), what is the voltage? What is the amount of work is done to move a charge from point a to point b?

$$\vec{\boldsymbol{E}} = x^3\boldsymbol{u}_x + 2x\boldsymbol{u}_y + z^2\boldsymbol{u}_z \tag{3.123}$$

3.5 What is the capacitance of a one meter long coaxial cable if the inner diameter is 1.5 mm and the outer is 4.5 mm?

3.6 Find the inductance for the previous problem. How much energy is stored in the inductive portion of the cable if 30 mA is sent through its DC?

3.7 Determine the power lost in a rectangular sheet of copper at 20°C if the sheet is 1mil thick (0.001 inches), 0.100 inches wide, and 1.3 inches long? If the copper is as wide and long as it is thick, compute the "ohms per square" for copper. How many squares fit into the entire length?

3.8 What is the wavelength for a plane wave traveling in a dielectric medium (only) with a permittivity constant of 4.2? What is the phase velocity? If a signal is traveling in the same medium at a frequency of 130 MHz and the medium is 1.5 feet long, how many wavelengths of the signal will reside in this medium?

3.9 Compute the skin-depth in copper for a 100 MHz signal. If the width is 1 mm, and length is 3 mm, how much power is lossed if the amplitude is 1 A?

4

POWER DISTRIBUTION NETWORK

The power distribution network PDN analysis has traditionally been the primary technical focus for power integrity engineers. It involves examining the impedance of the distribution path between the source and the load in the frequency domain. The reason for examining this network—rather than simply looking at the voltage droop at the package—is that critical signal behavior is unveiled through understanding the PDN, and often the analysis can be completed with minimal information from the silicon teams. The *efficacy* of the power signal from the load to the source is also an important factor in the integrity of the system power behavior. Additionally, from the amplitude and frequency of the key peak resonances in the impedance profile, insights could be gained as to where potential droop events could occur. Thus, usually the study consisted of examining the distribution path but occasionally including the load and sources in the system. (The load and the source for the system will be examined in later chapters.)

The distribution path contains power planes, parasitics, and sometimes a socket, a package, and the decoupling between the load and source. The source usually refers to the voltage regulator behavior and load behavior of the silicon. It is common practice to SPICE or the math-based program in analyzing this system to get a sense of its overall behavior. Actually, many good tools are available that can do this job, and the reader is encouraged to choose one that suits the circumstances and that they are familiar with. This chapter considers mainly the *concepts* that provide the basis for setting up and solving PDN problems. Initially, it will rely on *first principles* to show how they help at least in checking the results of simula-

Fundamentals of Power Integrity for Computer Platforms and Systems, First Edition. J. Ted DiBene II.

tions. The discussion will then break down the path, analyze its parts, put the parts together again to form a coherent PDN, and last present the methodologies used to analyze the PDN.

4.1 THE POWER DISTRIBUTION NETWORK

For a particular power path, the PDN comprises all of elements in the path that are critical for the proper function of the power delivery system. Key among these are the power converter, the power distribution network, and the load. For a high-performance PDN, there may many complex elements to consider. High-end computer platforms are examples of systems that have multi-faceted elements in the path, particularly in the power distribution network or the paths going to and from the silicon loads. By *complex*, this means that the geometries of the components and their extracted values are typically more difficult to analyze compared to other structures. To design the power distribution path in these systems requires a depth of understanding of the efficacy of the overall system power delivery. A firm grasp of the basics of the interconnect link that resides between the source of power and the load, is essential whether the interconnect is internal to the silicon device or on the printed circuit board, or the package.

The advanced silicon devices in electronic systems today all require high-quality power delivery. One of those devices is the microprocessor, which is used extensively in computer platforms; from handheld to large complex servers, they are considered the workhorse of the platform. In computer platforms that utilize high-performance microprocessors, not only are the voltages and associated tolerances on the printed circuit boards strict (in support of these devices) but the *dynamic* and *static* currents are typically very large, making management of the distribution path as much of a challenge as the design of the converter delivering the power [1].

Figure 4.1 gives an illustration of a two-socket server motherboard. Shown are some of the power converter components along the bottom of the figure, and two

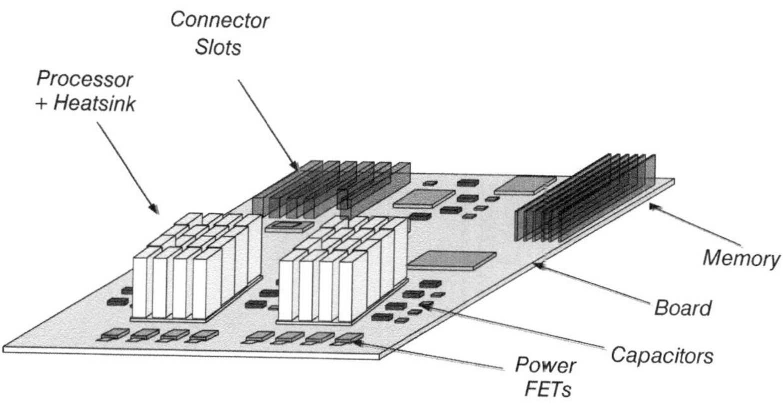

Figure 4.1 Two-socket server motherboard

high-performance microprocessors that reside under the heatsinks. Because today server motherboards usually carry a number of key components, such as slots for memory, additional slots for other *daughter* cards, a *controller* for the memory, and a *bridge* interface device, multiple voltage converters are required to power all these components, and the power delivered to them must be of relatively high quality [2].

Microprocessors also often have very challenging power delivery and distribution system designs due to the high dynamic and static currents that can occur during normal operation. When these devices are *multi-core* processors (more than one computing unit within a single silicon device), more than one voltage regulator may be required to power different functions of the chip [3]. The converters powering each portion of the device then require multiple *power* MOSFETs as well (large black devices at bottom of figure) as part of the conversion process to deliver the high currents (see Chapter 2). Each silicon device may use over a hundred amperes statically and switch at thousands of amperes per micro-second or faster [4]. Then multiple inductors and capacitors are required for energy storage and proper filtering for the subsystem link. Clearly, much forethought is required to properly lay out the plan and design the converter to ensure that the placement and routing of the power delivery and the choice of the power converter components meet the system power delivery needs. A first step (typically) is to translate the physical structure into a simple electrical representation that shows the characteristics of the distribution link, as in the following example.

Example 4.1 Determine the basic components *without values* in the distribution path of Figure 4.1. Assume simple lumped elements for each segment.

SOLUTION: The first step is to take note that there are five essential discrete sections to the distribution path: the voltage regulator, the PCB (printed circuit board) network from the VR to socket, the distribution under the socket and the socket interconnect—the package interconnect to the die—and the die interconnect and load. All this is shown in Figure 4.2.

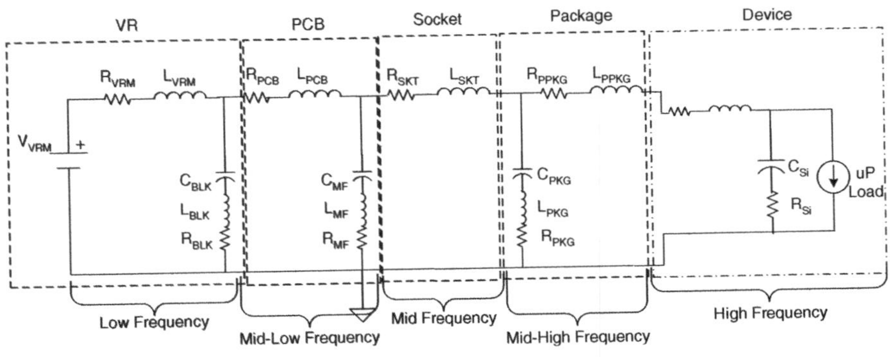

Figure 4.2 Typical power distribution path in complex power delivery system

So, to decode electrically the power path in Figure 4.1, now draw a simple electrical diagram that breaks down section by section the elements in the path. Going from left to right, the power source at the left is typically a DC-to-DC *Buck* converter that powers the system (as discussed in Chapter 2). This is the voltage source in its simplest form,[1] and includes some parasitic elements. The power converter's passive elements are usually directed toward the *low-frequency* filter operation. The voltage regulator (VR) has its own interconnect and is represented by a simple LR circuit along with the bulk capacitor network. The middle portion of the system is the main power delivery network or path. This is a passive network of resistors, inductors, and capacitors that physically connect and filter the power delivered between the source and the load. The *effective* frequency bands for these components range from relatively *low* to fairly *high* (see Table 4.1). This is the PCB PDN, the socket network, and the package. All these elements are analyzed separately or together to determine their effectiveness. The final section on the right-hand side of the diagram is the actual *load*. This is where the system connects internally to the silicon and externally to the load transistors (which are represented as a *lumped* current source in the figure[2]). There is also on-die decoupling, which is often represented by an RC network, often included in PDN analyses.

As power flows from left to right in Figure 4.2, the *effective* operational frequency band of the filtering increases. This is because the parasitics within each region (mainly filter capacitance and shunt elements) become less effective at delivering charge at higher frequencies as the power source moves away from the load. The typical frequency bands for each section are listed in Table 4.1. Note that these values are ranges and are not considered absolute.

As in the distribution shown in Figure 4.2, the frequency bands typically overlap due to the parasitics being continuous across the physical distribution path (series inductances and resistances). The capacitors contain their own parasitics and their operation is determined through their own impedance profiles (normally), usage

TABLE 4.1 Typical frequency range information on distribution impedances for high-performance PDNs

Impedance Region	Effective Frequency Band
VR	10 KHz–500 KHz
PCB	100 KHz–750 MHz
Socket	500 KHz–5 MHz
Package	1 MHz–15 MHz
Silicon device	10 MHz–100 MHz

[1]Note that the ground distribution is reflected into the voltage network. This is normally included in most PDNs but does not pertain to the details of the PDN. It is shown here only for completeness.
[2]Typically, this would be only one of multiple converters, which are required for a multi-core processsor.

(insertion parasitics), and type. Many of these capacitors reside on the *printed circuit board* and *package substrate* of the silicon device. The capacitors are intended to improve the *quality* of the power delivery path. It is usual for the engineer to separately analyze each section and compute the contributions from each—before, eventually, integrating the entire PDN to determine its overall impedance and efficacy.

Often it is desirable for the distribution to resemble (ideally) a *damped* transmission line impedance from the source to load [2]. This would allow the path to appear as a relatively *constant* impedance looking back from the active load at low frequencies and the resonances in the path to be *non-existent* at most of the operational frequencies to mitigate large voltage transients when the device is active and switches quickly [5]. Figure 4.3 shows schematically what this may look like. The frequency dependence is on the *resistive* portion of the network to ensure that high-frequency noise—if generated by the VR—is attenuated, but that the *low-frequency* components pass through. Additionally, from a step-load change perspective, it is desirable to pull the charge out as *fast* as possible from the nearest *charge source*, which is why the impedance is *lowest* near the load.

Figure 4.4 illustrates what this impedance profile would look like (lower panel) relative to the schematic. The impedance of the overall distribution rolls off at some frequency relative to the operation of the load to ensure that the relative voltage droop (ΔV) does not exceed the specifications. Thus, each section consists of a series *frequency-dependent* resistance, a series inductance, and a shunt capacitance. This means that the *reactive* portion is simply the *per-unit-length* LC portion of an ideal transmission line. As the frequency increases beyond a certain point, the *real* portion of the distribution decreases in value, and the overall impedance allows the charge to move into the load expeditiously as stated above. Though this approach increases the loss in the network, in the low-frequency band, droop events are more easily mitigated through the *lower impedance* of the PDN near the silicon load.

Though a damped structure is desired, and using lumped components certainly make the job easier, there are certain limitations in using *first principles* when attempting to get a basic understanding of this simple network. First of all, the impedance profile for a damped transmission line structure may not always look smooth when modeled—particularly if the segments are made up of rather large *lumped* elements. The lumped elements can create resonances that do not exist in the actual model. This can also result when modeling actual PDN circuits.

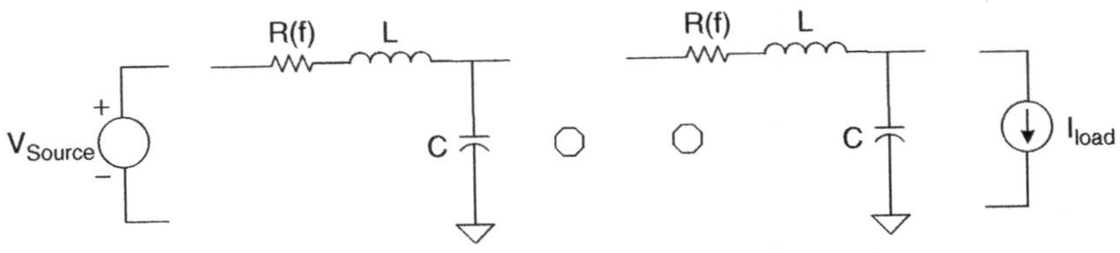

Figure 4.3 Simple model of damped power distribution network

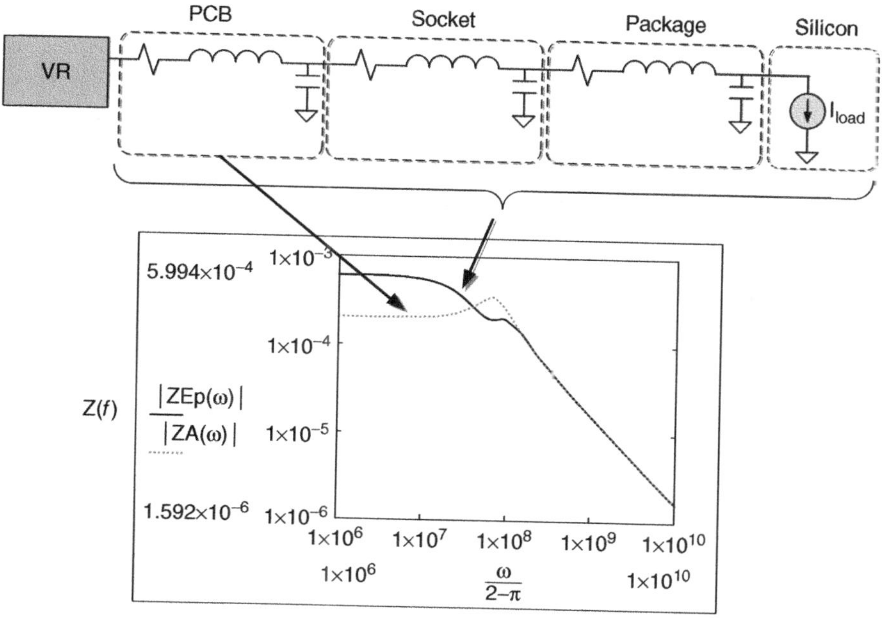

Figure 4.4 Example of an "ideal" power path

Care must be taken to ensure that enough segments are used to predict the behavior within the tolerance of the accuracy that is expected for the required modeling step. Figure 4.5 illustrates what the damped power distribution network looks like as the number of lumped segments are increased. This simple modeling exercise was done to emphasize such *resonant* behavior. As expected, resonance is seen for a *single* segment and the impedance begins to roll off on either side of the curve. This certainly shows that too few elements in the distribution path can cause a gross misrepresentation of the actual path. The next impedance plot in the figure has eight segments. Multiple resonances exist here but the distribution begins to *flatten* out slightly and shift toward the lower end of the spectrum due to the increased number of subsegments. The last profile shows a larger number of lumped elements and where the impedance is flat until the frequencies become very high. In many cases, the increase in impedance is expected to occur at a lower bandwidth point because of *material* behavioral (typically the thickness of various portions of the PDN in the copper planes) changes as the signals traverse the network. Note that the impedance increases at 10dB/decade in this example. This is the *skin-depth* effect on the actual thickness of the interconnect from the higher frequency currents, and it can be thicker than the actual skin-depth due to the high frequency nature of the switching currents in these PCBs and package layers. The ideal PDN would dampen only those frequency components that were unwanted and would transfer the power from source to load with little to no loss for the DC portion of the power signal.

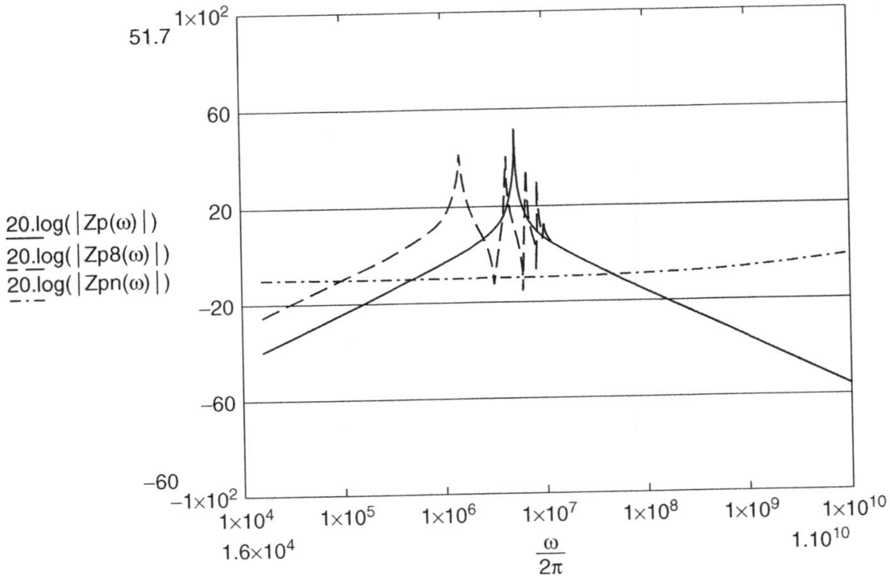

Figure 4.5 Impedance distribution of damped PDN for 1, 8, and 16 lumped segments

The bandwidth, or spectrum range for the PDN, is often well beyond the basic switching frequency of the loads and sources. A fairly *wide* spectral content of the impedance is examined because of this load device switching behavior. The harmonic distribution that is created by the load appears almost as *white noise* from DC into the GHz regime. The power integrity engineer has to allow for a controlled impedance distribution path—somewhat like a damped transmission line—because of this *wide frequency range* in the harmonic content of the load. In a world with minimal to no barriers, the PI engineer could design such a PDN structure in most cases [4]. However, in practice, this is extremely difficult to accomplish due to many factors—some of which are beyond the scope of this chapter. Nonetheless, this is a fundamental reason why the PI engineer and the system designers typically start with a physical layout and then analyze the distribution as presented to them by the silicon and package engineers. Once this network is understood both physically and electrically from the constraints given, the subsequent step is usually the analysis. The next section introduces the elements in the path and how to analyze them using *first principles* methods.

4.2 PDN ELEMENTS

Before putting together the full PDN for a high-performance network, from *first principles*, an examination of each of the main segments that make up this distribution, is warranted: the PCB, the socket, and the package distribution. Though many devices do not require a socket—and are soldered directly to the PCB—it is useful to become familiar with the effects from this element prior to looking at other package types. The simplist of these elements is the PCB.

4.2.1 PCB Network

The printed circuit board interconnect is the distribution that resides between the voltage regulator and the socket or chip package. Figure 4.2 shows a very simple lumped element distribution of this network. However, in actual practice, this part of the PDN is either simple or very complex depending on the application and design constraints in the system. In complex cases, *grid* models may be used to model the different portions of the network that make up the distribution. Typically, the PCB network traverses multiple power and ground planes before it ends up connecting to the socket.[3] This means that *vias* will be required in the path and at such connections, which can increase both the resistance and inductance of the path to the extent of affecting the PDN impedance profile. In Chapter 6, the segmentation method will be used to illustrate how multiple plane structures can be connected. In this chapter, only simple plane elements will be represented and modeled to help illustrate the methodologies.

Figure 4.6 shows a cross section and a plan view of a PCB interconnect where the current egresses out of the power converter and into the first planes. The layout engineer is usually working under a number of constraints due to the signal routing

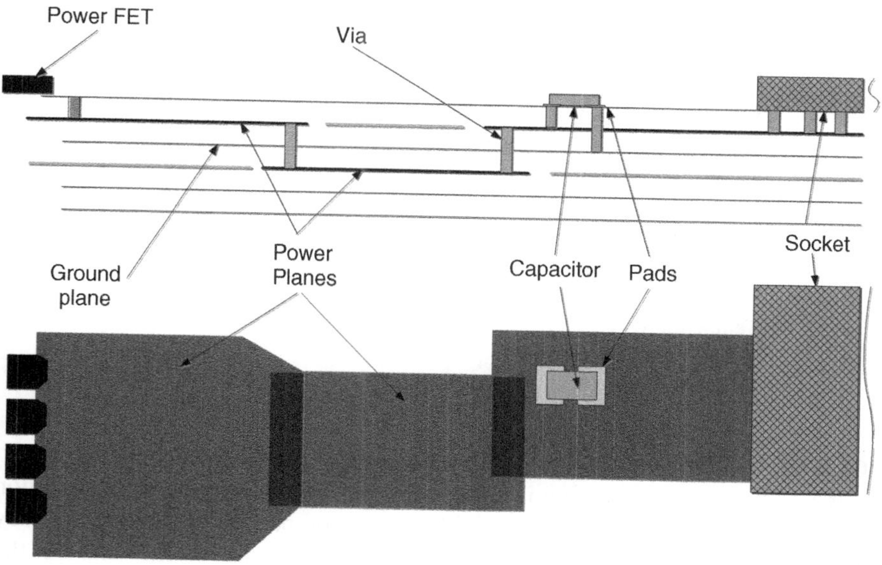

Figure 4.6 Simple PCB distribution cross section and "plan view" illustration

[3]The use of "socket" here is generic and may represent any connection that goes between the PCB and the package.

between high-performance devices on the board, so correct placement of the PDN layers can become a challenge. In today's high-performance computer platforms, the power planes are often routed first due to the stringency and complexity of the power distribution—particularly when multiple power rails are involved along with a (potentially) relatively large signal density.

In the model, each section, as in Figure 4.6, can have multiple elements and multiple components—mainly capacitors and their parasitics. The goal is to represent the salient elements in the path to gain an insight into its behavior. In the case of the figure, not all of the elements are shown since this is only for illustration purpose. The next step will be to represent a structure, such as the one in the figure in simple terms to allow one to model it correctly. The example below illustrates how to extract parasitics from a portion of this path.

Example 4.2 Using the data in Table 4.2, model the first plane on the far left-hand side of Figure 4.6 going into the plane to the right of the same figure. Determine both the inductance and the resistance using the *first principles* methods and equations from Chapter 3. Assume that the current distribution into the vias on the left- and right-hand sides are uniform—for example, assume that the inductor contact widths and vias into the plane are essentially uniform across the width of the plane itself. Further, assume the current follows into the taper on the right-hand side and drops down directly into vias at the edge (for simplicity) on the right side of the plane.

TABLE 4.2 Data for left and right computations

Plane/Pair Parameter	Metric
Length of nontapered section (end to edge)	3 in
Width of nontapered section (edge to edge)	1.5 in
Thickness (metal)	0.0014
Via rows (on left side)	3
Via spacing (row to row)	0.1 in
Dielectric thickness	0.008 in
Via (center to edge)	0.075 in
Temperature	40°C
Length of tapered section	0.5 in
Width of tapered section (smaller edge)	1.2 in

SOLUTION: The solution is easy if the computation is broken into two parts: first, for the portion of the plane that is *not* tapered and, second, for the tapered portion that is. Begin by using the data in Table 4.2 to compute the inductance. A key consideration is where to start and end the current distribution into and out the plane. If there are three rows of vias going into the first plane from the left-hand side (e.g., dropping down from the top surface into the plane), then the question is where should the inductance and resistance be measured? For a *first principles* estimate, the *average* point of the current *sink* is usually the best place for this, namely in the middle row of the *via* field. This is the place where the current will not flow into the via field equally in a practical design, therefore the best place to assume the end of the current distribution is where the *average* current would flow which should be closest in the middle via field. If the vias are spaced 0.1 inches apart, this will put the effective length of the plane at

$$Len_{eff} = Len - Via_{CtE} - 0.1 = 2.825\,(25.4) \cong 75.8 \text{ mm} \tag{4.1}$$

The inductance is estimated from equation (3.111):

$$L = \frac{\mu_0 ls}{w} = \frac{\left(4\pi e^{-7}\right)\left(75.8e^{-3}\right)\left(0.2e^{-3}\right)}{\left(38.1e^{-3}\right)} \cong 500 \text{ pH} \tag{4.2}$$

The resistance may be computed in a similar fashion; however, the temperature must be taken into account. The resistance changes in copper by 0.43% for every degree rise in temperature (approximately). Then, the resistance can be estimated as

$$R = \frac{2l}{A\sigma} = \frac{2\left(75.8e^{-3}\right)(1 + 20\,(0.0043))}{\left(0.036e^{-3}\right)\left(38.1e^{-3}\right)\left(5.7e^{7}\right)} \cong 2.1 \text{ m}\Omega \tag{4.3}$$

Note the factor of 2 for the resistance of both the power and the ground plane. The computations for the tapered sections are similar, except that the inductance and resistances are now simply the average of the two extremes of the widths of each section:

$$L = \frac{(4\pi e - 7)\left(12.7e^{-3}\right)\left(0.2e^{-3}\right)}{\left(34.3e^{-3}\right)} \cong 93 \text{ pH} \tag{4.4}$$

And the resistance

$$R = \frac{2\left(12.7e^{-3}\right)(1 + 20\,(0.0043))}{\left(0.036e^{-3}\right)\left(34.3e^{-3}\right)\left(5.7e^{7}\right)} \cong 0.39 \text{ m}\Omega \tag{4.5}$$

Therefore, the total inductance and resistance is

$$L_{tot} \cong 590 \text{ pH}, \; R_{tot} \cong 2.5 \text{ m}\Omega \tag{4.6}$$

Again, these are just approximations, but they can be a good first check of the models that will be later constructed for a more detailed simulation.

The PDN network for the PCB distribution can be made up of many elements—including a number of different capacitor structures. Refer, again, to Figure 4.6 for a moment. Note that each plane section has some amount of capacitance associated with it. This would be evident if the PI engineer were attempting to emulate a damped transmission line of sorts (or to just simply develop the best PDN possible). A first question that might be asked is, what type of capacitance and how much? This usually calls for an iterative approach to the problem, since often it is not known up front how much capacitance is appropriate to add to the PDN on the PCB without considering the entire problem space. For example, if the inductance and resistance of the planes are fairly large, the PI engineer might opt for putting a large amount of capacitance along the entire PDN until the socket is reached. The capacitor types could range from the larger values (and likely with larger parasitic Rs and Ls) on the VR side to the better, higher quality smaller capacitors near the socket. This would have the effect of reducing the large lumped element effects from the overall inductance of the path and thus help mitigate the overall droop from a large current step change. However, if the inductance and resistance are low, the engineer could use much less capacitance and opt for more high-frequency capacitance nearer the socket. The structure in Figure 4.6 is an example of a distribution that has both aspects to it. It is fairly long with multiple planes, but these planes are not high in inductance because of the relatively large widths and spacing of the planes—as in the example above. Thus, as a starting point, a number of capacitors could be added to the structure along the path. The physical structure would then look like the distribution network shown in Figure 4.7.

Depending on the number of capacitors required per plane for decoupling, the PDN may be represented by a series of lumped elements that may be connected in series. Note that, in the figure the ground path LR circuit has been reflected (once again) into the voltage side for simplicity. This can be done when the ground plane impedance is similar to that of the voltage plane; otherwise, it would be separated out into the return path. For most problems, even when the ground planes are very different, the network can be *averaged* to represent the total inductance and resistance of the network segment. The main reason for reflecting this portion of the network into the voltage side is to allow a simpler SPICE model to be created—at least initially. Often this representation is more than sufficient to get a reasonably accurate model of the interconnect for the *plane/pairs* for this portion of the PDN.

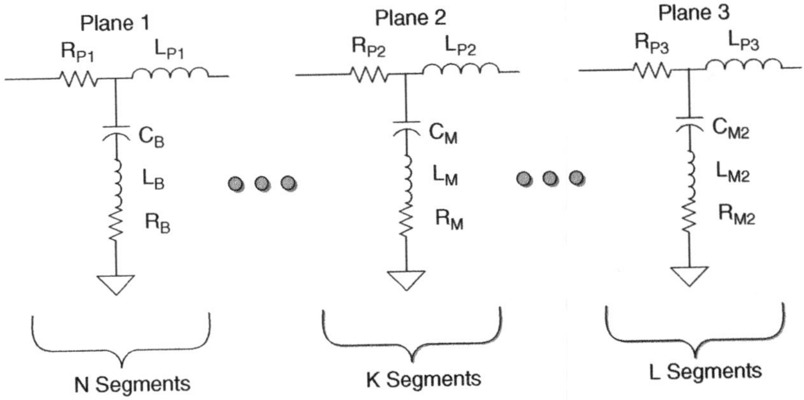

Figure 4.7 Lumped elements for the structure in Figure 4.6

Accuracy becomes more of an issue when the frequencies of interest are near the electrical wavelengths of the structure.

For plane 1, the LR network segment is a divisor of the values computed in Example 4.2. The capacitor is represented with its parasitics—since these will be used to determine the effectiveness of the decoupling from this section. As stated previously, the capacitors here normally have larger capacitance values and there are fewer of them—their purpose is to work with the local bulk decoupling for the LC filter of the converter and to limit any low- to mid-frequency noise in the range of 3 to 15 MHz. For plane 2, the LR network segment is broken down the same as in the first plane segment. But the capacitors are usually smaller here and have higher frequency resonances, or SRFs (series resonant frequencies), that allow them to perform better at higher frequencies than the bulk capacitors. The third plane segment again is close to the previous two, with the exception that the capacitors are of even higher frequency performance than those in the previous two plane-pair segments. This is consistent with the objective of trying to lower the impedance at higher frequencies as the electrical path moves closer to the silicon load.

It should be noted that how these plane regions are *divided* into smaller segments is dependent on where and how many capacitors are placed above the plane-pair. Figure 4.8 shows a breakdown example with a simple capacitor arrangement in the middle plane-pair. However, the capacitor structures are seldom placed symmetrically as shown, even parts that are staggered may be lumped together to form a segment that can be represented in SPICE or another program.

The two capacitors at the left side of the figure can be slightly different than the other two pairs to the right depending on (again) the PDN design. The two capacitors would then be lumped together with the actual plane/pair segment. This is what is called a *linear plane* model—meaning that each segment abuts linearly with the next segment in only one dimension. Later in the chapter, simple grid models will be discussed where the structures become more two-dimensional in nature.

The segments in Figure 4.8 can be broken down geometrically even though they are irregular in shape. In Chapter 6, other methods are discussed to combine these plane-pairs if more accuracy is needed—particularly at higher frequencies, and if other system aspects are of interest besides the power distribution behavior.

In Figure 4.9, a plot of the impedance for this simple plane is shown using the data in Example 4.2 and some capacitor values provided in Table 4.3. The curve starts out with the impedance matching the resistance of the segment from the example. Because the capacitors are reasonably large, the impedance is fairly *low* in the region around the resonance. Moreover, the large capacitance has a tendency to widen the resonance region. As the frequency increases, the inductance of the PDN begins to dominate, and the impedance increases around 20 dB per decade. Later, some simple tools will be used to analyze these types of structures more easily.

The next region in the PCB that is of interest are the *plane-pairs* that egress under the socket and/or package. This region is somewhat different from other PCB plane structures in that often the plane comes in from only *one* or two sides. What makes this unique is that the current distribution into this structure—both

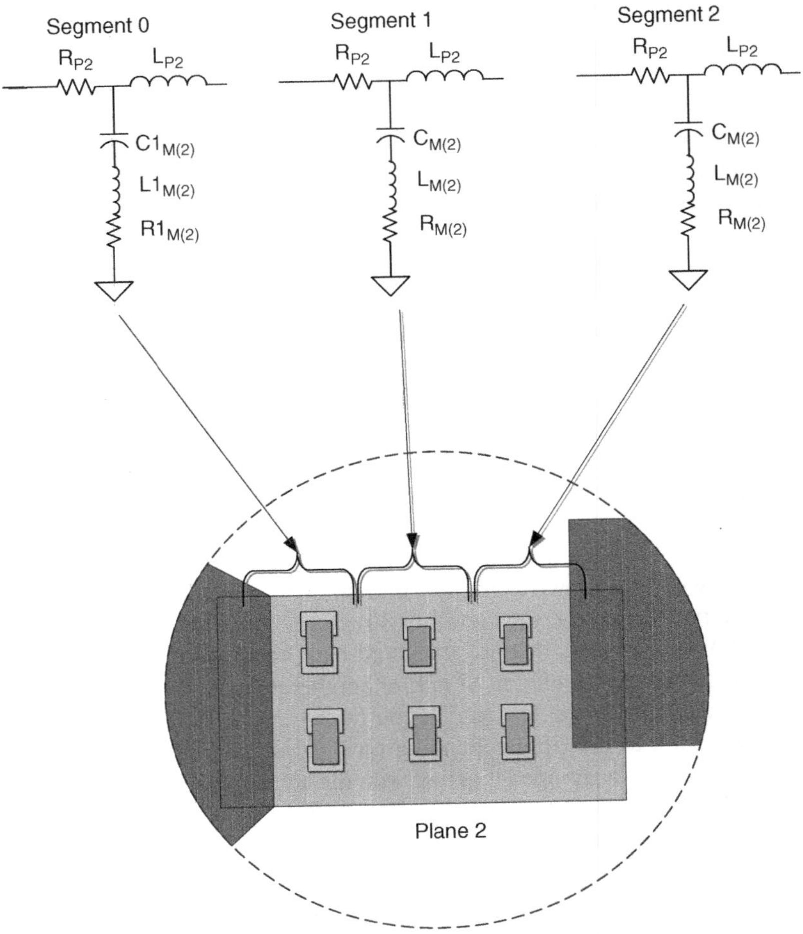

Figure 4.8 Plane 2 with capacitor and segment breakdown

AC and DC—is dependent on more than one factor. First, the *via* power/ground connections between the socket and the package/die location can have a very large effect on what the impedance of the distribution may be. This is because the path of *current* has a significant effect on both the inductance and the resistance of the connection—particularly, if there is little to no capacitance in the structure (for the mid- to high-frequency portion of the impedance). Figure 4.10 illustrates how potential current flow issues can arise if the via construction in the planes, socket connections, and packages are not designed with the load location in mind. To keep the impedance as low as possible, the current distribution is expected to flow as uniformly as possible through the copper that is available. If the vias are placed in a concentrated region, as shown in the figure, the current will also flow into that region, necking down into the via pattern. The via pattern, in this case, is showing a *left* to *right* flow into the corner of the socket. This is sometimes done to allow for

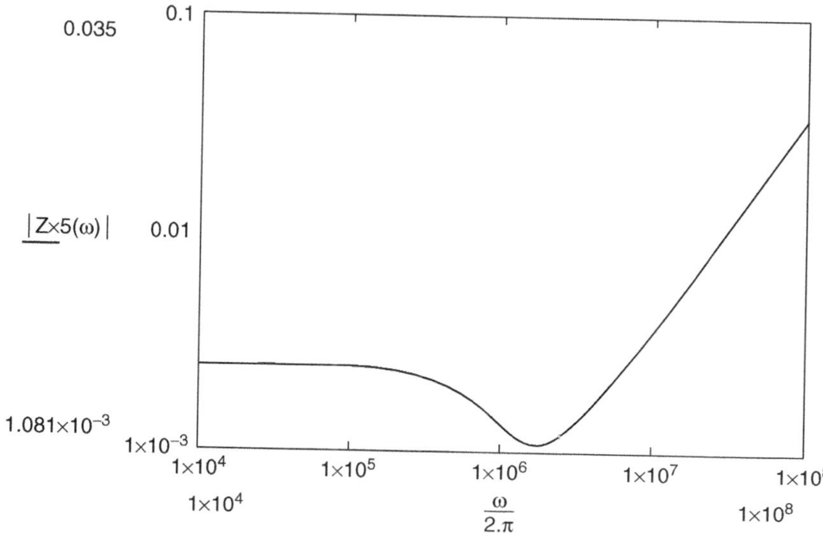

Figure 4.9 Impedance plot of single plane structure using Figure 4.8

TABLE 4.3 Capacitor data for plot

Capacitor Data	Metric
$C1_M$	47 μF
$L1_M$	400 pH
$R1_M$	5 mΩ
C_M	22 μF
L_M	400 pH
R_M	5 mΩ

other signals—grouping via and/or socket connections—and power to be routed (not shown) into the socket as well. Unfortunately, this has the effect of creating small lanes of copper between the vias, due to the drill patterns in the board, and limiting the flow of current into and out of that plane right underneath the socket. Because the power and ground planes are beneath each other, the plane nearest to the surface will have a larger *anti-pad* than the via diameter to allow separation from the other vias that have to pass through this plane. Additionally, the location of the actual load is nearly *orthogonal* to the current flow and is even *away* from other portions of the die near the location of the vias and socket connections. This will not only increase the impedance of connections to the final load but will also disrupt the flow of current into the other regions of the die.

When planning for the distribution network on the PCB, it is often best to work backward from the floorplan of the silicon—back into the socket and then into the planes of the PCB. This allows the board layout designer and power integrity

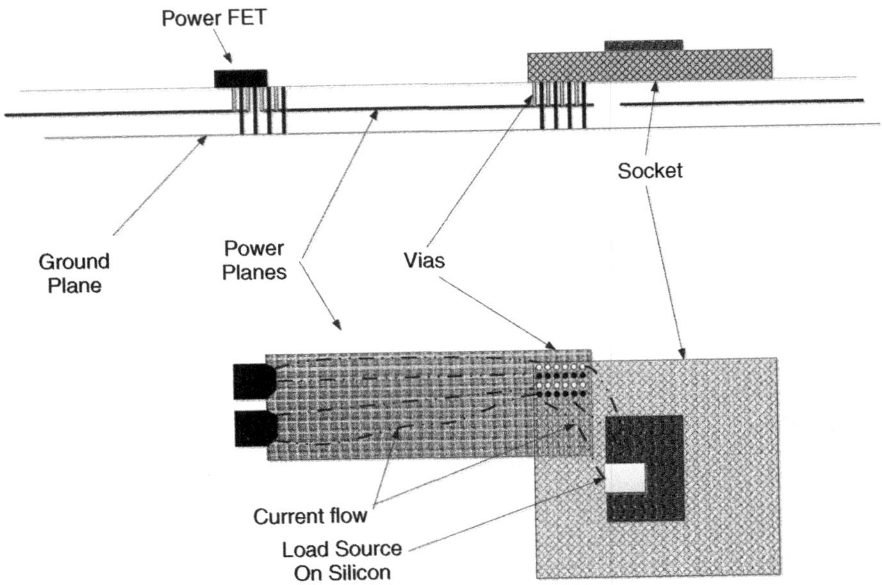

Figure 4.10 Current flow in poorly planned distribution

engineer to properly plan for the best PDN in the system rather than end up with a potentially unsatisfactory solution that compromises the electrical performance.

Later in the chapter, a structure such as the one shown in Figure 4.10 is illustrated with a better distribution and one that is more optimized for socket and contact conditions.

4.2.2 Socket Distribution

In the power distribution path there are always trade-offs between the selection of the connector, cost, area, number of pins, and other areas. It is not uncommon for the design engineer of the socket and package to overly constrain the PDN due to the limitations placed on their design—and these limitations can challenge the power integrity engineer. For example, the *pin map* position and/or bump contacts for the power distribution often conflicts with the signal IO, so there must be trade-offs between them. Therefore, it is recommended that the PI engineer be involved up front in the design process with the *pin-out* selection for the power delivery.

There are a number of parasitics that must be considered when analyzing systems with connectors. For power, the *pin* contacts may be represented by a combination of inductive, capacitive, and resistive elements. Typically, the resistance of the contact is what dominates this connection,[4] and the PI engineer needs to be aware of the

[4]There are examples where this is not the case. Some test sockets have connector systems that are more resistive than the contact resistance. For production level systems, though, the contact is usually dominant.

thermal issues associated with the connector as well as the position of the contact in the power path relative to the source and load. The capacitance of the socket pin is often ignored if the parasitics are used for power distribution because of their very small values. For the high speed I/O, the connector model may be more detailed and include multiple parasitics and electrical parameters.

The resistance of the connector involves not only the connector metal but also the contact, the latter often being the dominant part of the resistance and the biggest concern when analyzing the full path. Contact physics is a very complex field, and entire books and conferences[5] have been dedicated to its study [8]. The resistance of the connector and contact will only be touched upon here and in the next section. The reader is provided with some references at the end of the chapter for further examination.

Socket connectors for high-performance devices typically comprise LGA (land grid array) structures, though pin and socket connectors are still in use today. The purpose of using a socket instead of soldering the device onto the motherboard is that it gives customers the option of upgrading the device (provided that the new device is compatible with the old motherboard!). This is especially important for high-end systems such as servers and workstations where the cost to replace the platform may be prohibitably high for the customer.

The power bussed from the board to the package can be limited by electrical/thermal effects due to the position of the connector relative to the load and voltage source. The previous section discussed the electrical constraints on the performance of the PDN if the placement of the pin-out of the socket is not synergistically aligned with the layout of the printed circuit board. A poor electrical path can, of course, lead to system failures, but so can temperature increases due to current crowding into socket pins. [9].

Figure 4.11 shows a correct connector array distribution and distribution from the plane region throughout the socket path. The current flows from the planes directly under the via contacts to the LGA connectors. The arrays on the LGA contacts go directly upward into the planes of the package and then laterally into the die. The contacts for power and ground are arranged in a configuration that balances the current per contact in relation to the resistance of the path from one pin to the next. Contrast this picture with the one in Figure 4.10. If the socket contacts are not directly underneath the current flow, where some contacts may get more current than others, one or more power contacts may be susceptible to failure, primarily thermal runaway, as the current per contact hits its limit. The contact's resistance may also increase as a function of thermal, mechanical relaxation, age, and current density [9,11]. When the *number* of contacts are balanced, there are as many *ground* as *power* contacts. The current density per contact is not exceeded as long as the resistances of the planes are similar for the power and ground paths.

Based on the position of the load, all else being equal, the current will flow through the contacts nearest the power source. Thus, the nearest pins will take the

[5] The Holm conference is named for the famous engineer Ragnar Holm whose work in electric contacts helped to significantly advance the state-of-the art.

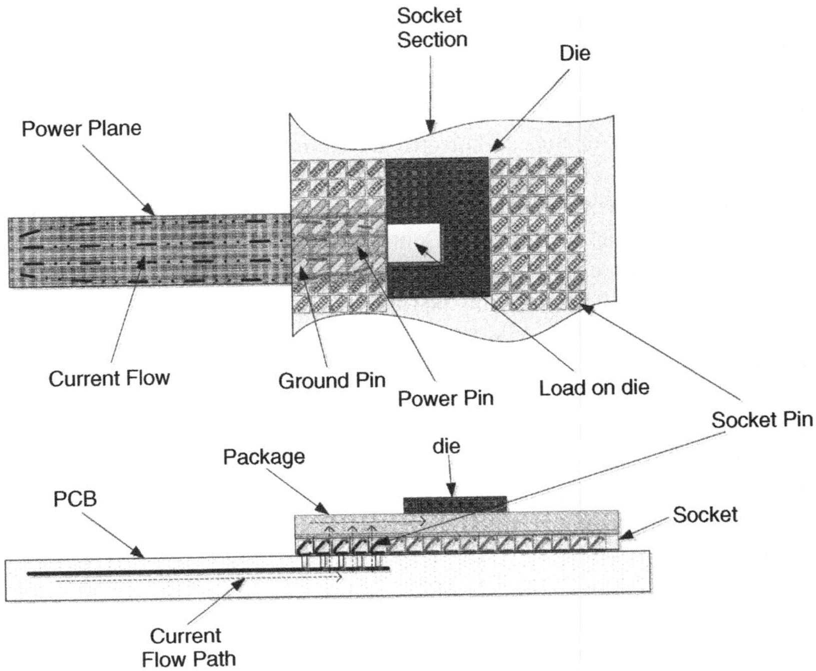

Figure 4.11 Top and side views of "good" PCB to load region layout

bulk of the current and heat up faster than the other pins. In the suboptimal case, a condition called *thermal runaway* occurs where the contacts open circuit and disconnect from the path. This puts the power delivery burden on the remaining contacts, and the aggregate current per contact increases until each contact fails in the system at a successively faster rate [12,14].

The resistance of the contact is particularly susceptible to many thermally and electrically related problems. Contact resistance is briefly discussed next with respect to power distribution and its impact on the PDN.

4.2.3 Contact Resistance

The LGA is the common choice for high-performance devices due to the performance and density requirements for these types of silicon. There are varieties of LGAs but most are *ball* mounted. The mounting of choice is usually a type of solder ball connection to the PCB. The connector's electrical characteristics (impedance), for the beam and contact point, is often represented by

$$Z(j\omega)_{Conn} = j\omega L_{beam} + \frac{1}{j\omega C_{beam}} + \left(R_{beam} + R_{contact}\right) \tag{4.7}$$

where the beam parasitics are distinguished from the contact. The beam includes the *solder joint* because it is part of the resistance path. Recall that, the resistance path

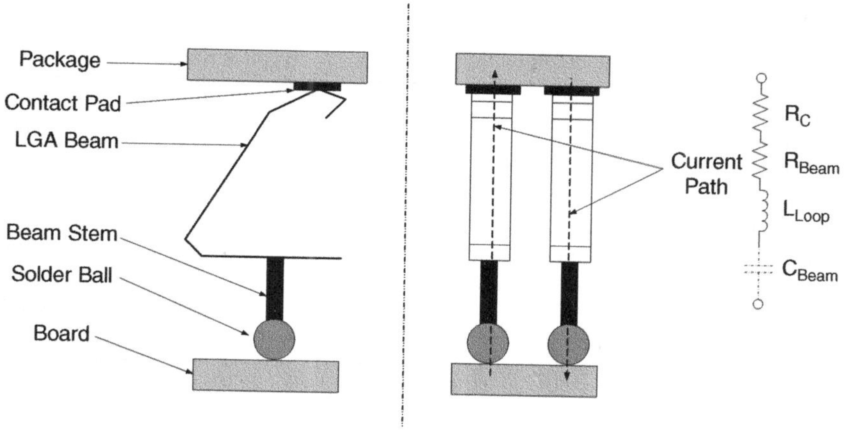

Figure 4.12 Side and front views of the LGA beam

is traversed across *both* the ground and power path. This is shown in Figure 4.12 in the front view.

Inductance, as one of the primary drivers for AC operation, is important to the analysis of the power distribution path. The capacitance of the connector is *typically* neglected for power integrity analysis but not in the signal integrity analysis for physical interfaces (as previously mentioned). The resistance of the beam may or may not be a major contributor, depending on the resistive contributions from the AC and/or DC components—the latter being largely dependent on the material properties.

Of the two resistances of the connector, the *beam* resistance can be estimated using *first principles* or via data from the manufacturer at a given temperature. The *first principles* computation is straightforward, and usually data on the material and the geometry allow for a reasonable estimate. The AC resistance, however, is more complicated and requires an additional analysis—if the frequency is well understood, an estimate based on the resistance due to the skin-effect will yield a reasonable data point. The last portion of the connector resistance is the *contact*. Estimating this value can be complicated.

Contact resistance usually involves estimating both *start-of-life* (SOL) and *end-of-life* (EOL) values. The physical properties of a contact are many and often quite complex. Table 4.4 shows some *typical* physical properties for a socket contact that is designed to deliver power between the source and load. All these properties affect the life span of the contact and must be considered when computing contact resistance—particularly for the *EOL* resistance value. Again the data points in the table are only estimates for LGA contacts, as there are many variations in LGA contact designs.[6] Indeed, for a great many contact types and structures, the data in Table 4.4 may not even apply for other contact and connector designs.

[6]It always good practice to check the data on any connector system with the vendor. This applies equally to LGA contact designs.

TABLE 4.4 Socket power contact characteristics and effects

Physical Properties for Contacts Providing Power	Typical Ranges	Unit
Contact resistance (SOL)	1–3	mΩ
Contact resistance (EOL)	1–15	mΩ
Contact normal force (SOL)	60–200	Mg
Contact normal force (EOL)	40–200	Mg
Base plating (typically Nickel)	50–100	μin
Contact plating (hard gold, dura gold, etc.)	15–50	μin
Contact wipe distance	0.25–1.0	mm
Contact wear	10–200	Cycles
Maximum operating temperature	120	°C

For both SOL and EOL resistance estimates, the system effects on the contact include environmental, wear cycling, and thermal aging. At start-of-life, the contact resistance may be computed in various ways. Some of the different methods are found in references [15–17]. *Holm* [8], for example, describes an approximate method based on *Hertzian* stresses on the contact surface, which is called the *constriction resistance* for the contact. Other methods that have been developed over the years allow for reasonable estimates for the start-of-life contact resistance—but, once again, these are only *estimates*.

To approximate the resistance at SOL, we will use a simplified formula found in [15]. At start of life, we can assume that the contribution from the *film resistance* is negligible; that is, there exists a *good* metal-to-metal contact interface. This is reasonable since the contact is, at that moment, free of debris and other contaminants that can create a film between the two metals at the interface. An *approximate* formula for the contact resistance is then

$$R_c = \frac{\rho}{\pi a} \arctan \frac{h}{a} \tag{4.8}$$

where ρ is the material resistivity, h is the constriction *depth*, and a is the radius of the contact area. Let us assume that the contact area is a *plane-sphere* of radius r (usually a good assumption). Then, the force that acts upon the structure, along with the modulus of the material, can be used to compute the resistance:

$$a = 1.1 \left(\frac{F_c r}{E} \right)^{1/3} \tag{4.9}$$

where F_c is the perpendicular force and E (not to be confused with the electric field) is Young's modulus for the material. In most contact systems, the ratio h/a goes to infinity. Combining equations (4.8) and (4.9) yields the simplified expression

$$R_c = \frac{\rho}{2.2} \left(\frac{E}{F_c r} \right)^{1/3} \tag{4.10}$$

Again, this is merely an *approximation* to the SOL resistance, but it can be used to predict expected measurements in the lab to gain insight into the resistance of the connector portion of the PDN.

End-of-life contact resistance is difficult to predict, and research on this subject can be found in a number of references—some are listed in the bibliography. EOL formulas based on a *linear interpolation* of the empirical data over accelerated life testing of the contacts exist in the literature. However, these data are application dependent and somewhat too speculative to use as a rule. It has been verified, though, that the effects on EOL resistance do include stress relaxation of the beam, film buildup, and fretting corrosion due to wear. Even these are technically hard to predict, either in part or together, particularly because of varying environmental conditions. For EOL predictions, it is best to consult with an electrical contact expert or the manufacturer—though, understandably such information from the manufacturer may not be easily forthcoming. The safest approach has been to predict a *worst case* EOL contact resistance and use this for all of the contacts in a socket though it is typically a poor method for predicting the accuracy in the actual contact resistance. The unfortunate consequence of this methodology is that it nearly always results in more contacts and a costlier solution than less conservative methods.

One *figure of merit* (FOM), for determining the EOL of the contact resistance of a contact—which may be used in conjunction with other data—is its resistance as a function of *force*. Figure 4.13 shows an example of this change over a *normalized* force curve. Note the differences at SOL and EOL for the same applied forces. The differences are due to many factors that make an accurate estimation (once again) difficult.

Curves such as the one shown in Figure 4.13 are often generated by the manufacturer through measurements of many connectors—where a statistical average is generated similar to that in the figure. Though usually *force* is one of the items that varies (stress relaxation), the two graphs show how SOL and EOL contact

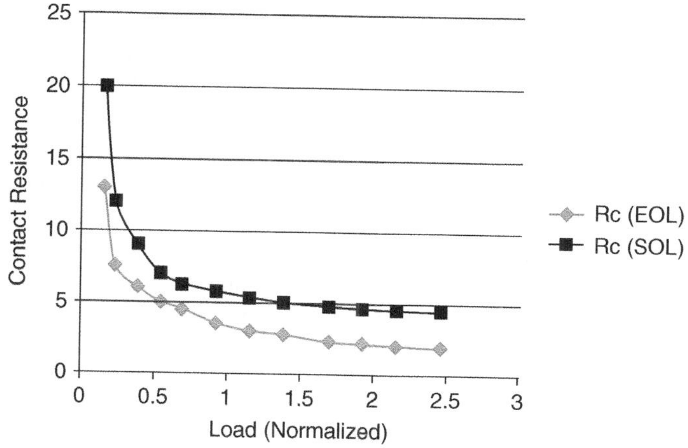

Figure 4.13 Graph of SOL and EOL contact resistance versus load

resistance can vary due to other factors. Movement of the contact due to vibration (*fretting effects*), and thus the formation of higher constriction resistance *A-spots* can be created, as well as film contamination. These and other effects can produce a larger EOL contact resistance for the EOL prediction—which is often used by the power integrity engineer as the contact resistance for the connector in the path. A problem at the end of the chapter illustrates a few of these issues.

4.2.4 Package Distribution

As discussed earlier in the chapter, current may be distributed into the package *differently* depending on where the power flows from. This distribution is moreover dependent on the layout of the package layers between the board connection and the silicon load device. An analysis of the package level power distribution primarily includes the package to board connections, the package interconnect (planes and traces), and the connections into the silicon. As seen in the previous section, sometimes a connector, such as a socket, is included. Because the filtering requirements are more stringent at the package level, noise is usually an important consideration as well. The power integrity analysis therefore must be more detailed than the board level analysis. Because of the higher frequency of the signals that are typically generated from the load die, and the proximity of the planes and capacitors to that load, the granularity of the analysis has to be *finer*, and placement of the capacitance (as well as the quantity) can further affect the results.

Up to this point, it has been assumed that the power and ground planes are the same for the power distribution and that one can simply *reflect* both paths into one circuit by duplicating the self-inductance or resistance of the voltage path. It has also been assumed that the current distribution through portions of the vertical path are essentially equal. It must be stated, once again, that in many, if not *all* designs, this assumption is not valid, particularly at the package level. Though, it is still an accepted practice to reflect the inductance and resistance of the ground path into the voltage path to simplify the analysis—particularly in SPICE—any *imbalance* between the voltage and ground nodes must be understood up front and accounted for to ensure that the correct parasitics are placed into the model. In addition, these parasitics must be placed within the proper *segments* of the model.

As in the PCB analysis, it is best to start by analyzing the package segment by segment. The first step is to examine the planes and the interconnects of the package from the capacitors down into the planes. For continuity, the socket and board are included to allow for an *exit* and *entry* path for the currents into and out of the package. Figure 4.14 shows a simple cross section for the package elements under consideration along with the schematic. (Chapter 7 provides a more detailed schematic that includes the die level parasitics.) Section A in Figure 4.14 shows the distribution from the socket to the package—the PCB and socket are illustrated here as well but only the socket is included in this distribution. Section B shows the *via* path up into the plane region of the package. This portion may egress into multiple planes, depending on which distribution region is within the design interest for the PDN. Section C is the first capacitor bank which in this case is the die-side

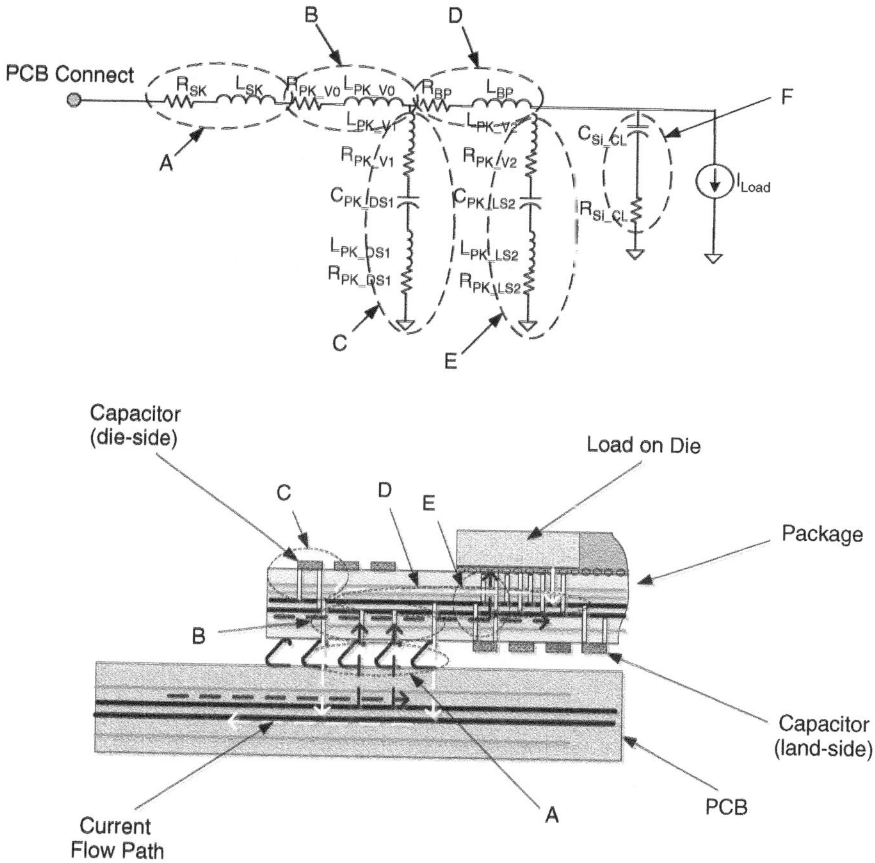

Figure 4.14 Package distribution example

capacitance. Depending on the design of the package, PI developers may put either *die-side* or *land-side* capacitors (or both) on the package for PDN decoupling. The land-side is often more effective due to the low impedance from the capacitors to the silicon (e.g., being a short vertical via route to the silicon device, typically). Section D is the distribution laterally from the die-side capacitors and/or the vias egressing into the main distribution planes across the package underneath the die. This may comprise more than one plane pair, but it is simplified here for illustration purposes to only *one* plane pair. Section E is the path from the capacitors on the land-side and the interconnect up to the silicon. If no land-side capacitors are present, this would only be an additional via parasitic from the planes where the currents would flow vertically. Finally, for completeness, section F shows a load die capacitor (simplified here to an R-C network) along with the load itself. For modeling many PDN structures, this is sufficient. If a PI engineer determines a large enough imbalance in the power/ground plane pair, both the inductance and resistance computations should take this into account. If an imbalance is evident from the geometrical differences in the planes, then the smaller plane usually is used as the baseline for the inductance computation (e.g., making the assumption that both planes are approximately the size of the smaller plane) and the resistance is computed as described in the previous sections into and out of the via regions.

Though it is not shown in Figure 4.14, there may also be capacitors on the PCB within the socket cavity region (assuming there is one) and on the backside of the PCB. Sometimes the socket cavity capacitance is included in the package model. This is because it is in the *shadow* of the package and the frequency response of the capacitors and the interconnect usually must be similar to that of the package decoupling.

As the network is traversed from the PCB to the package, the *quality* of the capacitors and their frequency response changes. The on-package decoupling is part of the decoupling for the high-frequency device load transients, which will occur in high-performance silicon devices. This decoupling is intended to have the *lowest* impedance at the highest frequencies to mitigate the high-frequency droops that occur due to large on-die load-step changes (see Chapter 5).

Typically, there are multiple networks associated with each section though only one impedance element is shown in Figure 4.14. Also the vias going from the capacitors to the inner package or PCB planes are a key part to the overall impedance of the capacitor, and so should be analyzed by the power integrity engineer. More often than not, however, the engineer assumes that the capacitor parameters are those of the specifications given by the vendor, but this is almost never the case. The *inserted* impedance of the capacitor is very different from what may be tested by the vendor due to the placement and routing differences between the two environments. Thus, it is usually up to the PI engineer to evaluate this as part of their analysis. More on capacitor types and technology is discussed later in the chapter.

The rate of transfer of the charge from the capacitor is dependent on the capacitor's placement (e.g., the parasitics of the connections to the planes and vias) relative to the interconnect path. This is different from the assumptions made in the previous section on the package distribution where it was assumed that the capacitors were mostly at an *equipotential* voltage. Even in a *first principles* analysis of the case in Figure 4.14, and the accompanying frequencies of interest, capacitor placement must be considered. Different types of decoupling capacitance are discussed in the next section. The next example makes use of the schematic and illustration in Figure 4.14 for a package structure.

Example 4.3 Develop a simple PDN model using the schematic in Figure 4.14 and the metrics in Table 4.5. Sweep the frequency from 100 KHz to over 300 MHz to observe any high-frequency affects, and discuss the resonances from the package, connector, and capacitors if possible. Include the socket in the computation.

SOLUTION: The parasitics have been computed already. If the geometries were known (or reasonable estimates made available), it would have been a simple task to put the formulas into a spreadsheet and compute them directly. Thus, it is just a

TABLE 4.5 Data for computations

Schematic Values	Metric
R_{SK}	1.4 mΩ
L_{SK}	42 pH
$R_{PK\_V0,\ V1,\ V2}$	42 μΩ
$L_{PK\_V0,\ V1,\ V2}$	5 pH
$C_{PK\_DS1\_LS2}$	30 μF
$L_{PK\_DS1,\ LS2}$	6.7 pH
$R_{PK\_DS1,\ LS2}$	300 μΩ
R_{BP}	1 mΩ
L_{BP}	293 pH
C_{Si_CL}	20 nF
R_{Si_CL}	400 μΩ

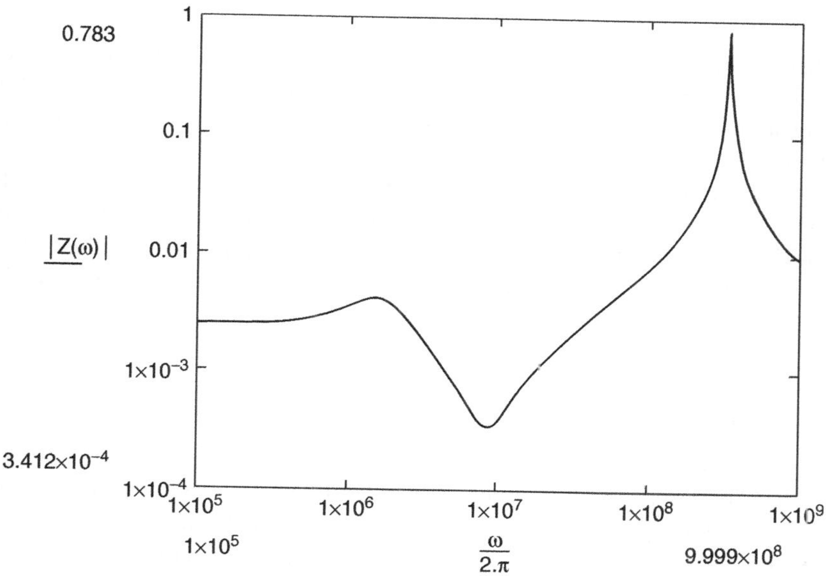

Figure 4.15 Impedance plot for the example

matter of putting together a PDN with a math program or SPICE. The value in using a Math program is that one can look at the equations and immediately determine what components dominate. Figure 4.15 shows the impedance plot. A good check as to whether the DC values are correct is to add up the resistances to see if they are reasonably close to what is shown in the plot at the low-frequency portion of the curve. This value is computed to be slightly over 3 mΩ, which is very close to what is observed on the plot.

The impedance plot in the figure has some interesting aspects to it. First, there is a small resonant peak around 1.6 MHz. This peak is likely due to the relatively large inductance of the connector/socket. This is expected, particularly if there are not enough contacts to lower this *effective* inductance. The second resonance appears to be due to the die-side and land-side capacitors.

Their SRFs, as will be seen later in the chapter in the discussion of capacitors, is within the impedance range where they are most effective—which is why the impedance *droops* here. The *high-frequency* impedance resonance is mainly due to the inductive parasitics of the planes in the package. This usually means that the impedance of the single plane-pair may not be low enough in this region. If a large step at the load should occur, it is likely this could cause a large high-frequency droop event (called first droop, as defined in the next chapter) that could cause the device to malfunction. Either adding high-quality on-die decoupling would help this (although often difficult) or changing the distribution planes in the package. Sometimes, getting better quality on-package capacitors can help (or more of them), but at some point the interconnect to these components becomes the dominant issue.

There are many different package types and designs, but most do not have sockets and are soldered directly to the board. For these packages, it is good to get the specifications directly from the manufacturer for the parasitics of the connections. Then, the PI engineer can usually work with the package designer to get information on the package traces or interconnects internally. These data are necessary for performing detailed parasitic extractions from a field solver. The engineer would start with a *first principles analysis* to get an idea of the actual PDN for this portion of the interconnect and then use the equations from the previous chapter (and other sources).

4.2.5 Decoupling Basics and Capacitors

The key to any good power distribution network is the *distributed* decoupling in the path. Good capacitor decoupling is essential to ensure that noise and droop is minimized in the path under the constraints of the load behavior. As one traverses along the distribution path, different capacitors are used for different portions of the link. Which capacitor, and its characteristics, is dependent on its location, the area available, and load characteristics that are required—either on the printed circuit board or on the package. The behavior of the capacitor is determined by the circuit elements of the capacitor and the parasitics of its connection when mounted. In the frequency domain, an important characteristic of the a capacitor—as mentioned previously—is its *series resonant frequency*, (SRF). This is the point in the impedance profile for the capacitor where its impedance is at its lowest. This point may be found by simply differentiating the equation for a capacitor

$$Z_c(j\omega) = R_C + j\omega L + \frac{1}{j\omega C} \tag{4.11}$$

and setting it to zero:

$$\frac{\partial}{\partial\omega}|Z(j\omega)| = \frac{\partial}{\partial\omega}\left|R_C + j\omega L + \frac{1}{j\omega C}\right| = 0 \tag{4.12}$$

Solving for the frequency yields the resonant frequency at which the lowest impedance point will be

$$f_r = \frac{1}{2\pi\sqrt{LC}} \tag{4.13}$$

Equation (4.11) is a simplified version of a more complex equation that incorporates many facets of the capacitance over the lifetime of the device. Some typical plots for the impedance of various capacitors (at start of life, or SOL) are shown in Figure 4.16. The lowest points on the plot indicate where the SRF of the capacitors occur. As shown in the figure, the SRFs increase from left to right as the quality of the capacitance (e.g., for high-frequency operation) increases. The capacitor with the lowest SRF is suited for *mid-* to *low*-frequency operation in the PDN and would typically be placed *close* to the MBVR. Conversely, the one with the highest SRF would be placed *closer* to the load and would be used for *mid-* to *high*-frequency operation. Normally, these are smaller in capacitance value but their inductive parasitics are better. Though size usually dictates the inductive portion of the structure, the quantity and placement of the power/ground terminals have a strong effect on inductance as well.

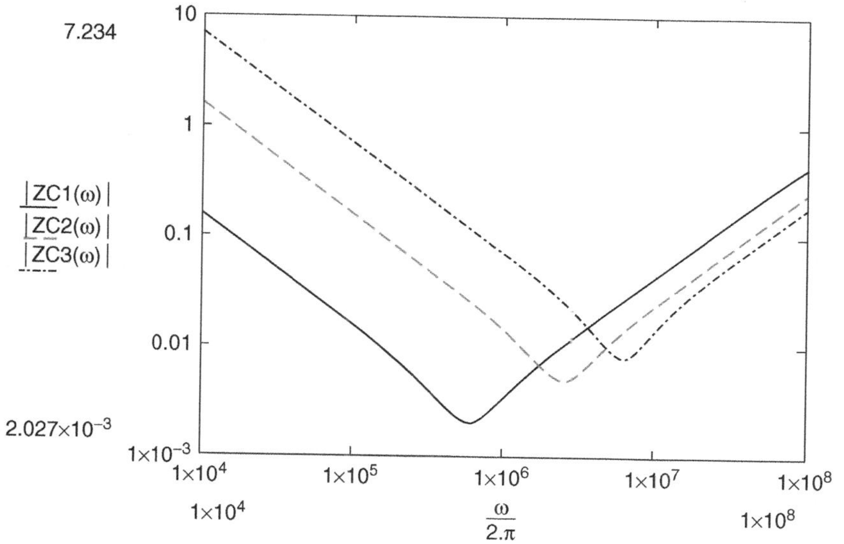

Figure 4.16 Impedances for various capacitor types, ZC1—low/mid frequency bulk; ZC2—mid frequency; and ZC3—mid/high frequency

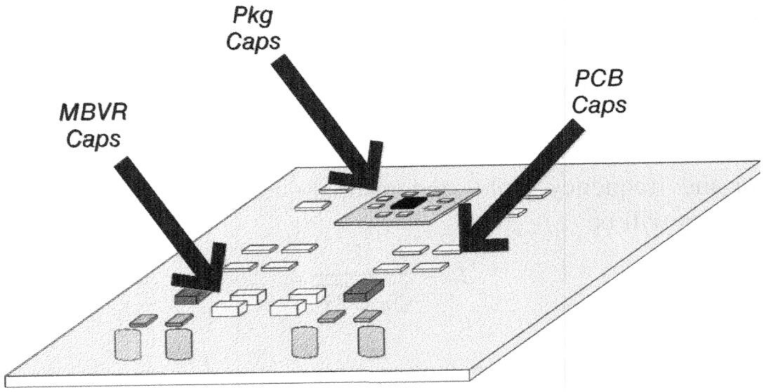

Figure 4.17 Typical capacitor position from voltage regulator to device load

The quantity and type of capacitors used for power decoupling vary depending on which portion of the power distribution path is being considered. Figure 4.17 illustrates the relative topology of some of these capacitors relative to the source (VR) and load (silicon device). In general, the position of the SRF point in the impedance curve *increases* as the capacitive networks get closer to the load device. This is because the objective is to *lower* the impedance in the frequency band of interest as one traverses the distribution network. Typical usages and types of capacitors for the PDN are shown in Table 4.6. As the frequency requirements for decoupling increase, the need for a higher quality capacitor increases, and thus the type, quantity, and performance changes.

A typical circuit diagram—as represented by equation (4.11)—is shown in Figure 4.18. This model usually suffices for capacitors where the impedance of the capacitance is *dwarfed* by the impedances of the parasitics. However, as the frequencies increase and other environmental factors come into play, portions of the capacitor may change in value. The power integrity engineer and designer need

TABLE 4.6 PDN capacitor types and usages

Capacitance Type	Capacitance Range (μF)	Typical Usage in PDN	Typical Frequency Range
Aluminum electrolytic	47–500	Input and output VR energy storage	10 KHz–1 MHz
Mid-frequency MLCC	0.47–200	Board level and device level decoupling	100 KHz–3 MHz
High-frequency MLCC	0.047–10	Package level decoupling and some board level	1–15 MHz
IDC (inter-digitated capacitors)	0.1–47	Package level decoupling	3–60 MHz

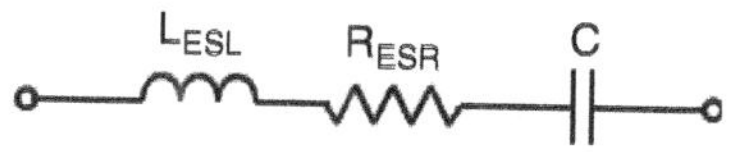

Figure 4.18 Simple capacitor circuit

to take such possible changes into regard, since they can affect the accuracy of the final distribution model and circuit.

For higher performance requirements, the electrical model for the capacitor may change significantly [18]. Circuits may need to be added to accommodate changes to the relative position of the current entry, mutual inductive, and capacitive coupling between plates, and conductive losses beyond plate resistance. For frequencies of operation below 100 MHz, which is typical for power distribution decoupling, the electrical parameters shown in the circuit in Figure 4.18 are usually adequate.

However, most PDN capacitance *does* vary with age, voltage rating, temperature, and dissipation. Thus, given these factors, the power integrity engineer needs to consider the proper values for capacitance when analyzing the network. The most commonly used capacitor type for the PDN is the *multi-layer ceramic chip capacitor* (MLCC) [19]. The MLCC has a simple construction that has proved for decades to be a very cost-effective and reliable capacitor for power decoupling. Figure 4.19 gives a simple physical representation of an MLCC package with internal layers. Usually multiple layers (often in excess of 50) are used to make up the total capacitance.

The total effective area of the capacitor can be found by examining the dimensions in Figure 4.19. Once the number of layers are known for each terminal, the total capacitance can be estimaed. A simple equation for a multiple parallel plate capacitor, as described in Chapter 3 and for the structure in Figure 4.19 is

$$C = \frac{(N-1)\,\epsilon A_e}{d} \tag{4.14}$$

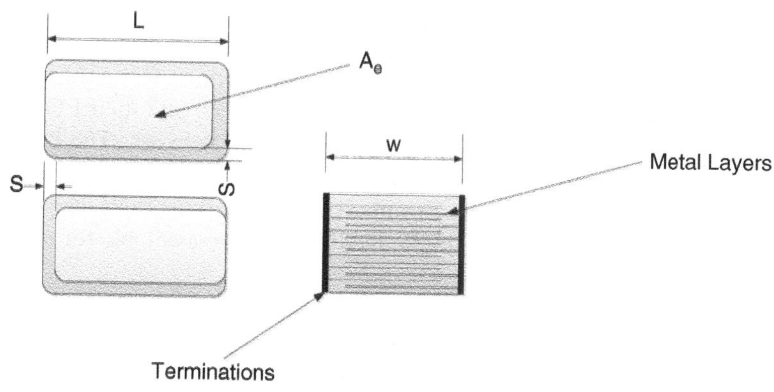

Figure 4.19 MLCC physical construction

where N is the number of layers, ε is the permittivity $(\varepsilon = \varepsilon_0 \varepsilon_r)$, A_e is the *effective* layer or *capacitance* area, and d is the thickness of a dielectric layer. As stated previously, over the life of the device, the capacitance may degrade significantly. These age effects are quite complex and often involve the modeling of electron trapping at the grain boundary [20]. The result is that the increased leakage current is similar to that found in nanometer transistor technologies [21].

Zhang et al. [20] determines age degradation through electron trapping based on the *current density profile* in the capacitor with the following equation:

$$J(t) = J_c(t) + E_0 \frac{\partial \epsilon'}{\partial t} - jE_0 \frac{\partial \epsilon''}{\partial t} \tag{4.15}$$

where $J_c(t)$ is the conduction current density. The permittivity ε is a complex quantity here and may be found by either using the *Kramer–Kronig* relation or assuming $J(t)$ to be known. The solution to the equation results in the following approximate relation to predict the age effects on capacitance for an MLCC capacitor:

$$C = C_0 - C_1 \ln(t) \tag{4.16}$$

where C_0 is the initial capacitance value and C_1 is a constant that is solved for. Equation (4.16) shows that the capacitance changes *logarithmically* with time. In addition, the displacement current I, varies as $1/t$:

$$I = \frac{VdC}{dt} \tag{4.17}$$

which is found by differentiating $Q = CV$ with respect to time, where the capacitance term is now *time dependent* and the voltage is assumed to be a constant (Chapter 3). It is not unusual for the capacitance to decay as much as 50% over the life of the device. However, multiple effects—including temperature and voltage—can also contribute to the decay. Figure 4.20 compares capacitance changes due to temperature for various capacitors. The y axis is the *relative*[7] capacitance in μF, and the x axis is in °C.

The power integrity engineer also needs to have on hand information on the effects of the capacitance changes over the life of the product. Often this information is available from data sheets on the various capacitors, and their EOL values thus can be predicted prior to analyzing the distribution network. The key in the prediction is to ensure that the proper ripple, system temperature, and expected life of the component is considered prior to embarking on a serious PDN analysis. The next section discusses some of the simple methods that are available for creating a PDN and then analyzing it.

[7]Relative here means normalized to a particular capacitor. The absolute values are not critical here; only the trends with respect to temperature. The curves in Figure 4.20 do not represent MLCCs in *general*, and the trends here do not represent *all* MLCCs. It is best to contact the actual manufacturer to acquire actual data on the specific capacitors one may use.

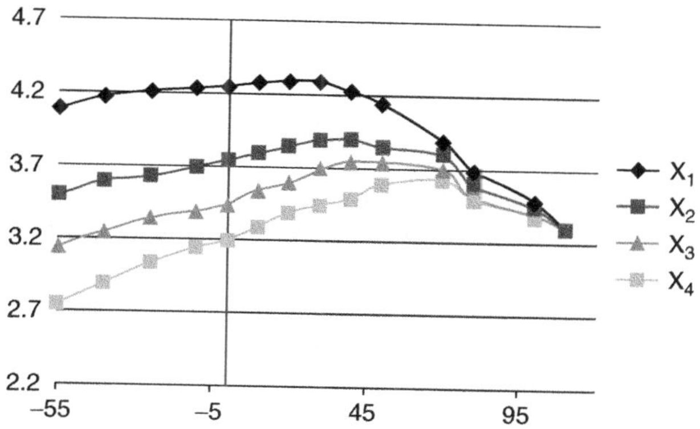

Figure 4.20 Plot of capacitance change over temperature for MLCC type capacitors, with X1 through X4 for the different capacitor types

4.3 IMPEDANCE DISTRIBUTION ANALYSIS

Voltage droop and the time-domain effects are important metrics for determining the efficacy of the power integrity of a PDN. Frequency domain analysis can give the power integrity engineer excellent insight into the network's behavior, so it is typically the initial check in the list of analyses a PI engineer must do. We have already considered in the earlier sections a number of simple plots generated for the PCB and package that show what the distribution(s) should look like. Now, we turn to simple *first principles* methods on how to generate these PDN structures mathematically.

4.3.1 Analysis of a PDN Structure through First Principals

It is usually good to start by re-introducing the PDN network with simple impedance networks, as shown in Figure 4.21.

The network in the figure is the familiar *ladder* structure that was encountered previously, but this time it is made up of simple *shunt* and *series* impedance elements. The total impedance of the network may be computed by looking back at

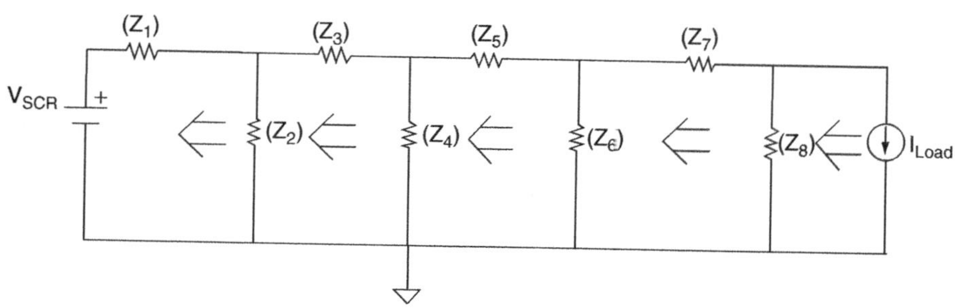

Figure 4.21 Impedance distribution of a generic PDN

the *load* from the *source*. The magnitude of this impedance can be written as,

$$|Z(j\omega)| = \left|\frac{V(j\omega)}{I(j\omega)}\right| \tag{4.18}$$

where the current and voltage are generally complex quantities and ω is the angular frequency of the signal, $Z(j\omega)$ is the total impedance, and $V(j\omega)$ and $I(j\omega)$ are the voltage and current, respectively. This is simply a *Fourier* representation of network impedance. The *phase* relationship of the plot is of lesser interest in a PDN analysis. Thus, the objective here is to determine the *magnitude* of the impedance network in order to examine the resonant behavior before attempting any complex analysis that would involve finding the phase between the voltages and currents. Each segment of the distribution path is represented by its own impedance network, which also comprises many elements as part of the overall analysis.

Figure 4.21 shows how the PDN can be broken down for easier analysis. Depending on which section is of interest, and looking back toward the voltage source, one would then determine the impedance distribution across the frequency spectrum section by section. Though the sections are not really independent, in the segmenting of the PDN more is revealed about resonant behavior than when the PDN is computed as a whole. For example, to analyze only the network looking back from Z_4, the equation would include only those circuits in the network path from Z_4 and below. Mathematically, this would be represented as a *nested fraction*:

$$|Z(j\omega)| = \left|\frac{Z_4(j\omega)}{1 + \dfrac{Z_4(j\omega)}{Z_3(j\omega) + \dfrac{Z_2(j\omega)}{1 + \frac{Z_2(j\omega)}{Z_1(j\omega)}}}}\right| \tag{4.19}$$

This equation is derived by looking at the impedance from each section individually and then computing the series and parallel impedances in a *nested* fashion. Equation (4.19) can be simplified by dividing through by Z_4 and Z_2 at different parts of the fraction,

$$|Z(j\omega)| = \left|\frac{1}{\dfrac{1}{Z_4(j\omega)} + \dfrac{1}{Z_3(j\omega) + \frac{1}{\frac{1}{Z_2(j\omega)} + \frac{1}{Z_1(j\omega)}}}}\right| \tag{4.20}$$

to obtain a *continuous fraction*. Then, the total impedance can be computed by inspection depending on where one is looking back from within the network. Moreover, various math programs are available that can readily do the simplification of nesting each fraction from bottom to the top. In the example of Figure 4.21, the series and parallel impedance segments would be placed into the fraction by identifying their placement relative to the continuous fraction. For the first three segments, the individual impedances would be computed between the nodes (e.g., Z_1 between the VR and the node connecting to Z_2 and Z_3). In the case of the impedance looking back into Z_4 this would be computed iteratively using the following equations:

$$Z^{i1} = \frac{Z_1 Z_2}{Z_1 + Z_2} \tag{4.21}$$

$$Z^{i2} = Z^{i1} + Z_3 \tag{4.22}$$

$$Z^{i3} = \frac{Z^{i2} Z_4}{Z^{i2} + Z_4} \tag{4.23}$$

By the iterative approach any errors in the computation can be easily detected, since each equation can be checked independently. If there is further need to *roll-up* the continuous fraction into only one fraction for a whole PDN analysis, equation (4.20) can be rewritten as

$$|Z_{tot}| = \left| \frac{Z_4 \left[Z_3 \left(Z_1 + Z_2\right) + Z_1 Z_2\right]}{\left[\left(Z_3 + Z_4\right)\left(Z_1 + Z_2\right) + Z_1 Z_2\right]} \right| \tag{4.24}$$

where the individual function variables have been omitted for simplicity (as in the previous equations). To determine the distribution impedance, which is done by generating an asymptotic *Bode* plot of the impedance distribution [24], equation (4.24) is transformed, for ease of computation, as

$$|Z_{tot}| = C_n \left| \frac{\left(1 + \frac{K_1}{j\omega}\right)\left(1 + \frac{K_2}{j\omega}\right) \cdots \left(1 + \frac{K_n}{j\omega}\right)}{\left(1 + \frac{L_1}{j\omega}\right)\left(1 + \frac{L_2}{j\omega}\right) \cdots \left(1 + \frac{L_n}{j\omega}\right)} \right| \tag{4.25}$$

where K_i, and L_i are the *break frequencies* of the impedance distribution and C_n is a constant. Then, from the log of equation (4.25), a form that is more conducive to plotting is taking:

$$\log|Z_{tot}| = \log|C_n| + \log\left|\left(1 + \frac{K_1}{j\omega}\right)\right| - \log\left|\left(1 + \frac{L_1}{j\omega}\right)\right| \cdots$$
$$+ \log\left|\left(1 + \frac{K_n}{j\omega}\right)\right| - \log\left|\left(1 + \frac{L_n}{j\omega}\right)\right| \tag{4.26}$$

The next example illustrates how equation (4.26) can be used to generate a simplified impedance profile.

Example 4.4 Using Figure 4.21 and equation (4.26), determine the magnitude of the impedance of the last two segments in the figure (VR portion mainly) in terms of the components below and then approximate the impedances of the network at DC and at very high frequencies (above 1 GHz) given the following values:

$R_{VRM} = 1\ m\Omega$

$L_{vrm} = 200\ \mu H,$

$R_{blk} = 100\ \mu\Omega,$

$L_{blk} = 200\ pH,$

$C_{blk} = 2000\ \mu F$

SOLUTION: The impedance for the VR segment is the form of equation (4.24) is

$$|Z_{pcb\_VR}| = \left|\frac{[Z_1 Z_2]}{[(Z_1 + Z_2)]}\right| \tag{4.27}$$

The equation is then put into the form of equation (4.26), by which the break frequencies are determined, and from there the Bode plot is created. The key operation is to plug in the given values and then use a root solver to find the frequencies. The actual zero-pole equation, in terms of the variables, is left as a problem for the reader at the end of the chapter. The values for the constants are plugged in, and equation (4.27) is simplified for the break frequencies:

$$|Z_{pcb\_VR}| = \left(1e^{-3}\right)\frac{\left(1-\frac{\omega}{5}\right)\left(1-\frac{\omega}{50e^{-3}}\right)^2}{\left(1-\frac{\omega}{1575}\right)^2} \tag{4.28}$$

where $\omega_i = \omega_1, \omega_2, \omega_3 = 5{,}1575{,}50{,}000$. The equation for the Bode plot representation takes the form

$$db|Z_{pcb\_VR}| = -60 + 20\ \log\left(1-\frac{\omega}{5}\right)$$
$$-40\ \log\left(1-\frac{\omega}{50e^{-3}}\right) + 40\ \log\left(1-\frac{\omega}{1575}\right) \tag{4.29}$$

The graph can now be generated using the given break frequencies. The gain is −60 dB down base on the DC value for R_{vrm}. The plotted function is shown in the asymptotic Bode diagram of Figure 4.22. Note the break frequencies in the diagram. The actual impedance plot is shown in Figure 4.23.

As expected, the two plots align fairly well, particularly at the break frequencies. The amplitudes are off due to the discontinuities in the computations of the asymptotes, which is to be expected. The value in using this type of analysis is that

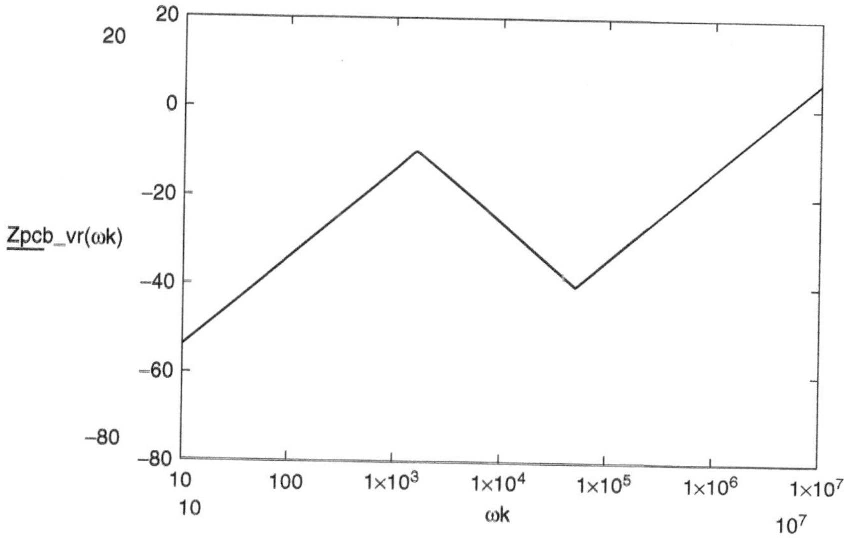

Figure 4.22 Asymptotic Bode plot for Example 4.4

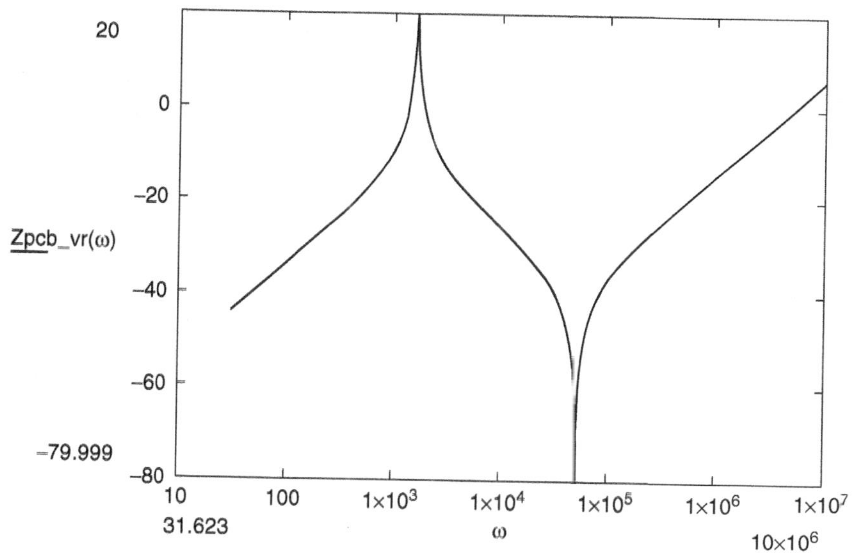

Figure 4.23 Actual impedance plot for example

it allows an engineer to estimate (relatively) where the resonances are in a PDN without having to do extensive modeling. Using equations (4.25) and (4.26) for a moment allows one to determine the high-frequency and low-frequency behavior of the circuit. When the frequency is close to zero, the impedance is 1 mΩ. This is the series resistance of the path as is predicted by the equation. When the frequency is very high, the equation predicts that the impedance will rise at 20 dB per decade. This is also expected because the inductive parasitics begin to dominate above a certain frequency.

However, using *nested fractions* for every PDN analysis may not be practical when the number of nodes becomes significant. For larger problems it is often more judicious to use SPICE or some other program to perform the number crunching. Even for larger, complex problems, performing a simple "sanity" check on the analysis is always a good idea.

There are other methods that allow for PDN modeling when the planes are larger—as in the PCB—and involve more segments and granularity [23]. *Grid models* use subsegments that allow designers to analyze the effects of a PDN for distributions that spread over so large an area that a simple ladder network would suffice [25]. Figure 4.24 gives an example of a grid model where a plane is segmented into smaller grids to allow a PI engineer to gain insight into the behavior of the larger structure, but also uses tools such as SPICE to simulate the PDN. In the figure, each segment, or *cell*, represents a portion of the plane. Wherever a capacitor is placed in the plane, the network for this capacitor is joined—as shown in the lower image—to a junction nearest one of the segments.

The size of the network is typically dictated by the highest frequency of interest and is often sized according to a fraction of the wavelength of the signal. For example, if the highest frequency of interest is up to 500 MHz, and the dielectric constant is 4.2, a safe segment size would depend on the propagation of that signal within the plane-pair for the PDN, or mathematically (from equation (3.97) of chapter 3),

$$l_{size} \leq \frac{\lambda}{8} = \frac{v}{8f} = \frac{300e^{6}}{8 \times 500e^{6}\sqrt{4.2}} \cong 3.7\ \text{cm} \tag{4.30}$$

which is relatively large compared with the actual dimensions of the cell that one might use for a PDN. This result from equation 4.30 ensures that there will not be a resonant behavior within the cell at the highest frequency of interest.

Additionally, grid models work fairly well with parasitics extracted from field solvers, and grid models allow for very detailed SPICE models to be built when a complex model is required.

4.3.2 Analysis of a Full PDN Structure

After examining different segments and elements in the PDN, the next step is to put the results together to determine what the entire PDN would look like and how it could be analyzed. The analysis can be simplified by combining segments where appropriate. Again, a simple ladder model can be used. Problem 4.5 shows the values used in the schematic of Figure 4.25—the proof of the impedance structure is left as an exercise for the reader to verify that the same PDN was achieved as well.

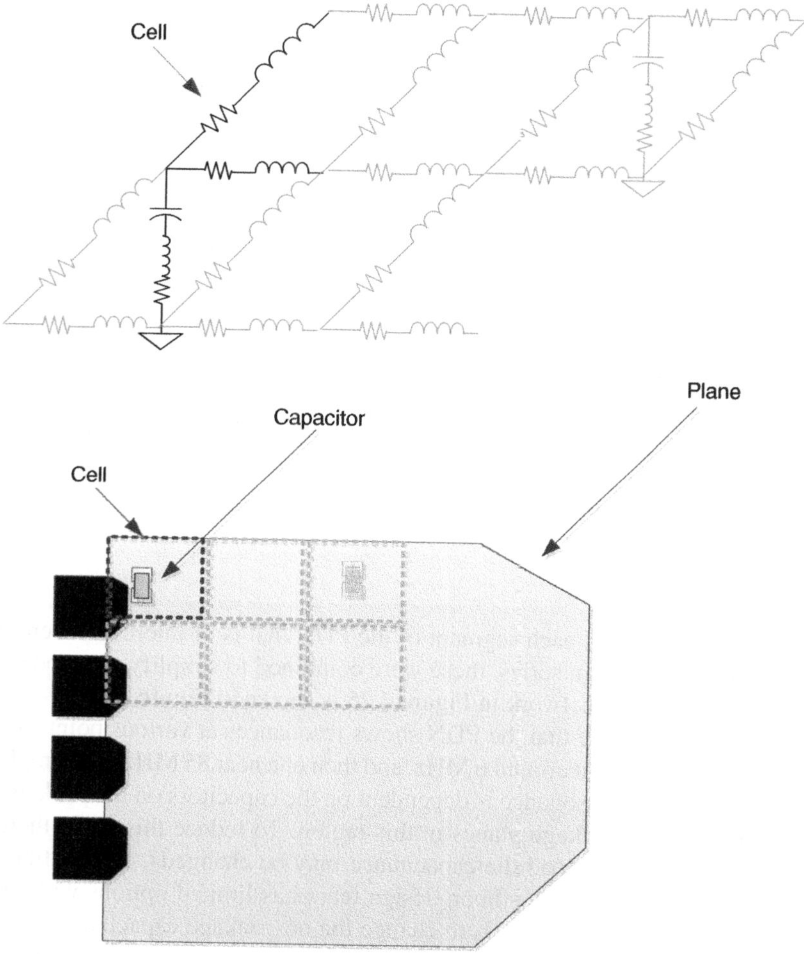

Figure 4.24 Cell(s) of a grid model representation

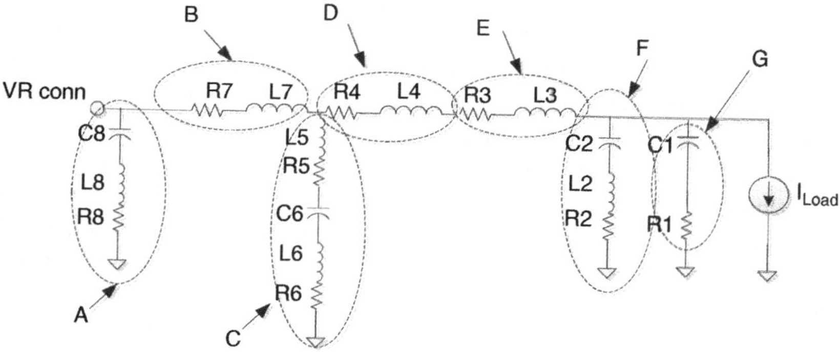

Figure 4.25 Schematic for full PDN analysis (simplified)

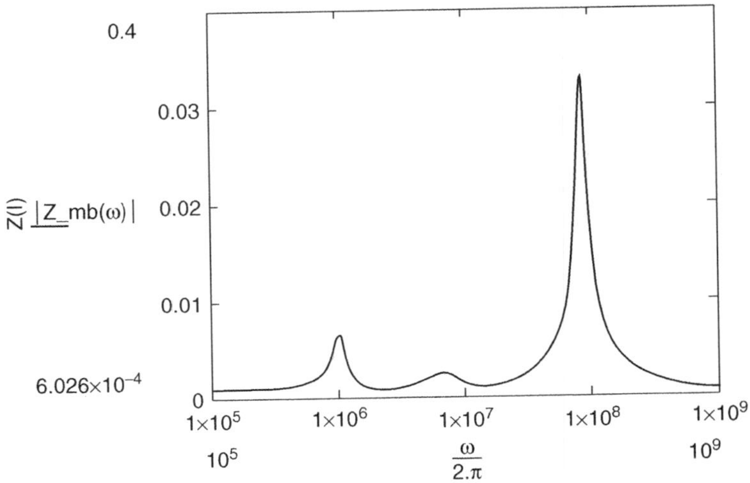

Figure 4.26 PDN for schematic in Figure 4.25

The schematic shows each segment of the PDN that is of interest. Where there were lumped segments in series, these were combined to simplify the model. The impedance plot for the network in Figure 4.25 is shown in Figure 4.26.

In Figure 4.26, notice that the PDN shows resonances at various points: one at around 1 MHz, another at around 6 MHz, and then one near 85 MHz. The amplitude of the high frequency resonance is dependent on the capacitors on the package and the parasitics of the package planes in this region. To reduce this peak, there are a few options. First, the load-die capacitance may be changed—more often than not, this is difficult since the silicon design team has limited options with on-die capacitance. The second option is to change the on-package capacitance. Though less difficult, usually there are constraints here as well—particularly on quantity and placement. Often the first question that arises is what is the cost and impact? The impact is measured by the voltage droop, which is discussed in detail in the next chapter. The cost side of the equation is dependent on the first issue. If the specifications are met through both the frequency-domain and time-domain, then usually there is no need to reduce the resonances. In fact, sometimes managers ask if capacitors can be removed to save cost [26]! Third, if the package or die cannot be changed, the final option is to try and affect the resonance from the PCB. This is by far the most challenging since the impedance from the board through the socket is usually too high to lower by simply changing the capacitance on the board. This is why it is best to work very closely with the system design team(s) up front to ensure that the PDN design is fully comprehended in the development cycle.

In many PDN designs, the question of the effectiveness of the capacitance is first and foremost. The next example illustrates in a PDN how capacitance can affect the result.

Example 4.5 A board designer has developed a power delivery system for a high-performance silicon device and wants to determine how many capacitors can be removed from the distribution network on the printed circuit board to reduce cost and save board space [26]. He wants a first-pass estimate before he asks for a detailed analysis from the PI engineer. The impedance needs to be below 2 mΩ from 100 KHz to 10 MHz, since the loop response of his VR is at least 100 KHz. Given the values for the network components below, compute the impedance of the network using the methods from this chapter—looking back from the socket and determine the profile up to at least 100 MHz. Include the impedance of the VR portion from earlier in this section.

Values

$R_{PCB\_1} = 0.2\ \text{m}\Omega$

$L_{PCB\_1} = 10\ \text{pH}$

$R_{PCB\_2} = 0.1\ \text{m}\Omega$

$L_{PCB\_2} = 5\ \text{pH}$

$R_{PCB\_3} = 0.4\ \text{m}\Omega$

$L_{PCB\_3} = 20\ \text{pH}$

$R_B = 400\ \mu\Omega$

$L_B = 40\ \text{pH}$

$C_B = 400\ \mu\text{F}$

$R_M = 100\ \mu\Omega$

$L_M = 10\ \text{pH}$

$C_M = 50\ \mu\text{F}$

SOLUTION: Here the objective is to determine the effects on changing the capacitors. The examination will be to look at the impedance profile for three cases: all of the caps, half of the capacitors, and then no capacitors. The first step is to look at the circuit by inspection and determine the continuous fraction for the network.

The impedance of the network may be simplified to the one shown in Figure 4.27. Note that, once again, the resistance and inductance of the return path have been reflected into the voltage side of the network. This is acceptable only if the values are similar or can be averaged without significant loss in accuracy (as discussed previously).

Each equation can now be put into a Mathcad sheet, and a *nested analysis* can be applied to determine the effective impedance of the structure; then, the values for all of the impedances can be put into each impedance as a variable. This involves simply computing the value for the first fraction and then using that value to solve the next fraction, and so on.

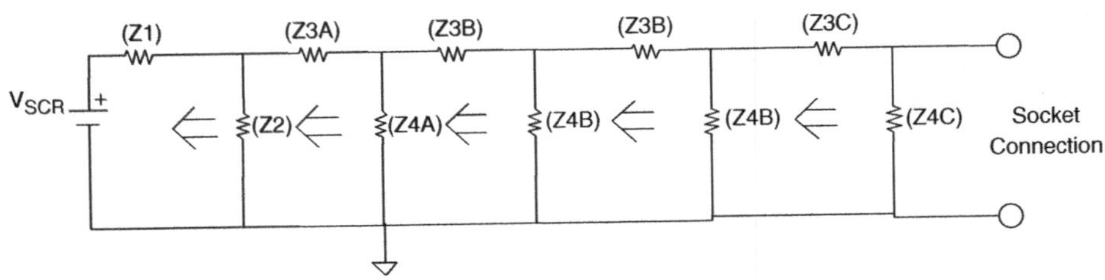

Figure 4.27 Simplified schematic for analysis

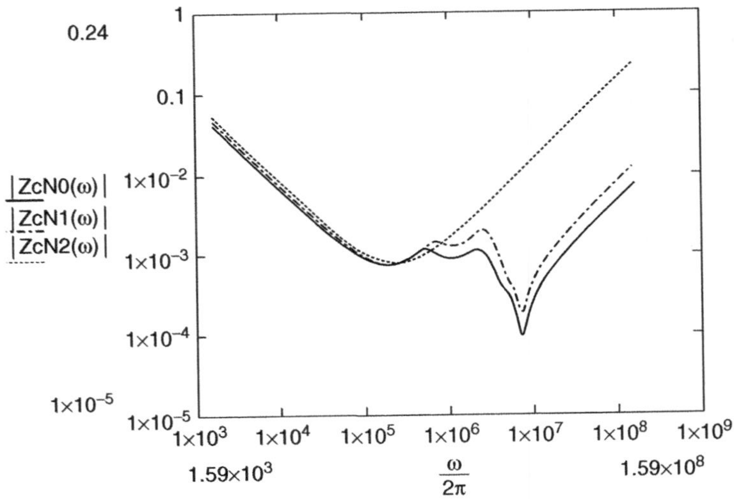

Figure 4.28 Impedance plot with three different values of capacitance in place

The plot in Figure 4.28 shows the full results with the three variations of capacitance added. Here both axes have been plotted in a log scale to show the frequency and amplitude behavior more precisely. The low-frequency resonance that appears is mainly due to the bulk capacitance of the VR. Though it is quite low, it is well within the bandwidth of the switching converter, and thus, anything below the loop response of the converter should be ignored for now, since the VR will respond to that. The impedance, however, does roll off fairly slowly, and near 100 KHz it is close to 1 mΩ. The curves are very similar in this frequency regime, as expected, since they are dominated by the VR bulk values. The more interesting aspect is the effects of decreasing the board capacitance from one-half to zero. This can be seen from comparing the *bowl*-shaped curve with the middle one in the figure. From about 300 KHz to 10 MHz the lack of capacitance is evident, and this begins to indicate that the impedance will increase quite significantly on the PCB. Thus, unless one is able to address this problem with additional capacitance on the package, it appears the prudent thing to do is to add some amount of capacitance to keep the impedance below the required value of 1.3 mΩ and lower the noise path.

Many power integrity problems are solved by making assumptions and recognizing when a simplification of the problem can yield satisfactory results. More often than not, that the PI engineer is faced with deciding whether a simple model will be adequate for the analysis and whether this work concludes with a reasonable solution. It is consequently a matter of experience and resources that guides the process. Unfortunately, all too often the engineer begins with the details and then has to revert to a simplified view of the problem. As discussed in Chapter 1, the objective of this book is to help engineers make judicious choices when delving into such complex problems.

Additionally, before embarking on the PDN modeling effort, a few rules-of-tumb have proved useful:

- A *first principles analysis* can reduce the number of errors and iterations in the modeling effort. It is usually expeditious and gives real world insights into the problem.
- A field solver is handy for checking the values for each parasitic element. If the *first principle* equations are reasonable but the numerical result is off by more than 50%, it is a good idea to recheck the numerical model for error. Most errors are found in the boundary conditions. Sometimes the error is as simple as the placement of a ground point that yields larger or smaller loop inductances than actually exist in the physical structure.
- Check for resonances in the results that may not be there. Examine the highest frequency and also the final SPICE model (or a similar aid) to verify that there are enough elements to predict the behavior correctly.
- Always have another pair of eyes examine the work. Many hours or days of extra effort can be reduced by simply asking one colleague check the work of another.

4.4 SUMMARY

The ideal PDN should resemble a *damped transmission line* and have some decay to it. The impedance of a PDN is a *complex* quantity, with magnitude and phase. Typically, the PI engineer is only interested in the *magnitude* of the impedance distribution. The PDN is often broken down into simple sections: the VR portion, the PCB region, the socket or connection to the PCB, the package, and the die region. Each section can often be simplified initially using simple ladder networks—where the components of each is reflected into the voltage region for analysis in SPICE or a math program. The PCB sections can be broken up into plane-pairs and analyzed together or separately with capacitors placed within each structure. Connectors to packages, or sockets, require an understanding of the impacts from both the inductance and resistance—the dominant effect being the contact resistance. The package is often analyzed in more detail because of the high frequency effects in the path and its proximity to the load device. Capacitors play an important role in reducing the resonances in the PDN and keeping the impedance flat throughout the operational bandwidth. The *SRF* of the capacitor is a good starting metric for verifying the quality of the capacitor. Capacitor behavior changes as a function of age, temperature, and voltage, and in most cases, poor behavior will affect the final solution. A common tool is to use a *nested fraction* to determine the impedance network through *first principles*. The key resonant points may be found by using a Bode plot analysis.

REFERENCES

1. Gupta, M. S., et al. Understanding voltage variations in chip multiprocessors using a distributed power-delivery network. Design, Automation and Test in Europe Conf. and Exhibition, 2007.
2. Hochanson, D. H., DiBene II, J. T. *Powerdelivery for high performance microprocessor packages—Part I*. Tutorial Presentation: IEEE EMC Symp; 2007.
3. Sinkar, A. A. Workload-aware voltage regulator optimization for power efficient multi-core processors. IEEE Design, Automation and Test in Europe Conf., pp. 1134–1137, 2012.
4. Novak, I., Smith, L., and Roy, T. Low impedance power planes with self damping. IEEE EPEP Conf., 2000.
5. Wang, J. S.-H, and Dai, W. W.-M. Optimal design of self-damped lossy transmission lines for multi-chip modules. IEEE Int. Conf. Computer Design VLSI in Computers and Processors, 1994.
6. Ren, Y., Yao, K., Xu, M., Lee, F. C. Analysis of the power delivery path from the 12-V VR to the microprocessor. *IEEE Trans. Power Electronics*, 2004.
7. Selli, G., Drewniak, J. L., DuBroff, R. E., Fan, J., Knighten, J. L., Smith, N. W., Archambeautl, B., Grivet-Talocia, S., Canavero, F. Complex power distribution network investigation using SPICE based extraction from first principal formulations. *IEEE EPEP*, 2005.
8. Holm, R. *Electric Contacts: Theory and Applications*, 4th ed. Springer, 1967.
9. Rosen, G., Coghill, A., and Tunca, N. Prediction of connectors long term performance from accelerated thermal aging tests. *IEEE Electrical Contacts, Proc. 38th IEEE Holm Conf.*, pp. 257–263, 1992.
10. Queffelec, J. L., Ben Jamaa, N., and Travers, D. Materials and contact shape studies for automobile connector development. *Holm Conference Proc.*, 1990.
11. Malucci, R. D. The effects of wipe on contact resistance of aged surfaces. Holm Conf., pp. 131–144, 1994.
12. Figueroa, D. G., Chung, C. Y., Cornelius, M. D., Yew, T. G., Li, Y-L. High performance socket characterization technique for microprocessors. EPEP, 1999.

13. Mroczkowski, R. S. Concerning reliability modeling of connectors. Holm Conf., 1998.
14. Rosen, G., Coghill, A., and Tunca, N. Prediction of connectors long term performance from acclerated thermal aging tests. Holm Conf., 1992.
15. Caven Jr., R. W., and Jalali, J. Predicting the contact resistance distribution of electrical contacts by modeling the contact interface. Holm Conf., 1991.
16. Glossbrenner, E. W. The life and times of an A-spot. Holm Conf., 1993.
17. Malucci, R. D. Multispot model of contacts based on surface features. Holm Conf., 1990.
18. Murphy, A. T., and Young, F. J. High frequency performance of capacitors. *IEEE ECTC*, 1991.
19. Rostamzadeh, C., Canavero, F., Kashefi, F., Darbandi, M. Effectiveness of multilayer ceramic capacitors. *IEEE in Compliance*, 2013 Annual Guide, 2013.
20. Zhang, T, Yoo, I. K., and Burton, L. C. Low-frequency measurements and modeling of MLC capacitors, *IEEE Trans. Components, Hybrids, and Manufacturing Technology*. 1989.
21. Narendar, S., et al. Full-chip subthreshold leakage power prediction and reduction techniques for Sub-0.18 μm CMOS. *IEEE J. Solid-State Circuits*, 2004.
22. Brown, R.W. Distributed circuit modeling of multilayer capacitor parameters related to the metal film layer. *IEEE Trans. Components and Packaging Technologies*, vol. 30 2007.
23. Shi, H., Oatley, J. L., Joseph, R., Wei, G-Y., Brooks, D. M. An experimental procedure for characterizing interconnects to the DC power bus on a multilayer printed circuit board. *IEEE Trans. EMC*, 1997.
24. Van Valkenburg, M. E. *Network Analysis, 3rd ed.* PrenticeHall, 1974.
25. Mezhiba, A. V., and Friedman, E. G. Electrical characteristics of multi-layer power distribution grids. *IEEE ISCAS*, 2003.
26. Fizesan, R., and Pitica, D. Simulation of power integrity to design a PCB for optimum cost. IEEE Int Symp. Design and Technology in Electronic Packaging (SIITME), 2010.

PROBLEMS

4.1 Create a network from the L-R values in Example 4.2 and the capacitor data in Table 4.3. Put together two plots: one has one lumped segment and the other is made up of 8 segments—where the segments are divided by 8 in a distributed fashion. Sweep the plot from 100 KHz to 1 GHz. Is there any difference in the impedance plots? Did you expect any?

4.2 A PDN requires that the socket have a SOL contact resistance of less than 2 mΩ in order to meet the test requirements for overall validation. Assuming that the force is 65 g per contact and Young's modulus is 78×10^{10} N-m, the A-spot radii is around 300 μm, and the conductivity is that of hard-gold, estimate the contact resistance at SOL. Compare the result with that in Figure 4.13 if the normalized force value is also 65 g.

TABLE 4.7 Data for Problem 4.5

Schematic Values	Metric
C8	1 mF
L8	6 pH
R8	50 μΩ
L7	214 pH
R7	216 μΩ
C6	94 μF
L6	2 pH
R6	50 μΩ
L5	30 pH
R5	30 μΩ
L4	11 pH
R4	360 μΩ
L3	9.2 pH
R3	236 μΩ
C2	10 μF
L2	16 pH
R2	1 mΩ
C1	300 nF
R1	600 μΩ

4.3 There are 35 pairs of contacts in the arrangement in a socket. Determine both the SOL and EOL resistances for the connector PDN portion using Figure 4.13 using the contact resistance computed in Problem 4.2.

4.4 Determine the generic zero-pole equation for the impedance function using the variables in Example 4.4.

4.5 Verify the PDN in Section 4.3.2 using the values in Table 4.7.

5

POWER INTEGRITY TIME-DOMAIN AND BOUNDARY ANALYSIS

In Chapter 4, the components of the power distribution network were reviewed and then the impedances were analyzed to determine where resonances occur across the frequency bands. However, the question of what comprises a *good* performing network was left unanswered. In this chapter, the time-domain effects of such distributions are addressed. Voltage droop behavior within these networks is also examined along with the boundaries of the voltage droop.

Generating the network components and learning how to analyze them is an important first step in understanding the PDN. This chapter broadens the circuit structure to include the loads and sources in addition to the PDN. Thus, the attention is turned to the time-domain along with the introduction of some basic modeling tools that help the PI engineer analyze problems in both the transient and steady-state regions. The chapter begins with the basics on how to model the sources and loads in a system and then turns to examining the behavior of the networks in the time-domain. This type of analysis lends itself toward using programs such as SPICE and MathCad for circuit and mathematical modeling, respectively. Therefore, many of the examples and analyses will be making use of such tools. The reader should have a reasonable understanding of how to use SPICE [1]. In this book, PSPICE [2,3] is used for circuit modeling.

5.1 SOURCE AND LOAD MODELING

In Chapter 2, various power converter topologies were introduced to show how power is delivered into the network. For simple PDN analyses, it is usually

Fundamentals of Power Integrity for Computer Platforms and Systems, First Edition. J. Ted DiBene II.

cumbersome—and rarely necessary—to provide a *sophisticated* switching model of the source in order to determine the efficacy of the PDN. Rather, it is more efficient to represent the source in terms that complement the PDN without distorting the results, since the objective is not to determine the *quality* of the power supply but rather the quality of the *network* under various operating conditions created by the load. Thus, when representing the power converter for these analyses (which, for simplicity, will not include converter losses), simple circuit elements and behavioral models will be used to create a representative view of the source.

The load representations are a different matter. For time-domain simulations, the current source most closely represents the system's PDN transient behavior. The silicon team typically gives the parameters of this current source, but an understanding of how it is constructed and where it comes from is important. Thus, the first objective will be to investigate the current's perturbations in order to find the boundaries and the operational limits of the PDN and system power network behavior.

5.1.1 Source Representations

When modeling power converters as sources, there are a number of trade-offs that must be considered. First, the accuracy and complexity of the model should be understood up front. For many systems, a simple ideal source is sufficient—for others, a closed-loop switching model is used. Second, there are usually time constraints that limit the PI engineers' ability to deliver the results within a given schedule. The issue is typically not the complexity of the model, though that can become a problem at some point, it is the *iterative process* that follows where the engineer's energy is usually most expended. Thus, part of this consideration is the expected accuracy of the results from the first step of the process. Finally, the relative tools that are available become an important part of the decision process. Most engineers have access to some type of SPICE analytical tool set and a math program, such as Matlab or Mathcad. When doing the modeling, it is good to know the limits of the tool sets. Some analytical methods are included here that allow the PI engineer to use a simple Excel spreadsheet and/or just pen and paper.

There are different power converters for computer platforms that are chosen based on the power source. A few were discussed in Chapter 2. Given that most computer systems use *buck converters* as their power source, the focus will be on a simple representation of this topology throughout the discussion. Once again, the objective is to model only enough behavior to allow the results of the time-domain analyses to be correctly represented. Figure 5.1 shows a simple *switched* schematic for a single-phase buck converter.

For many time-domain analyses, the PI engineer will want to ensure that the converter *behavior* allows representative results for the time-domain simulations. Thus, the PI engineer will need to understand some basic features of the converter:

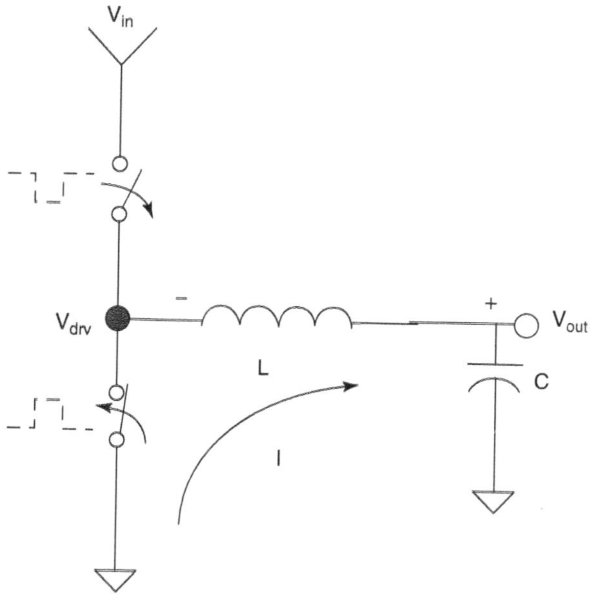

Figure 5.1 Topology of an ideal buck converter

- The open loop bandwidth
- The amount of filter capacitance
- The switching frequency

The open-loop response determines how fast the converter will respond to a change in voltage without feedback. Typically, the *open-loop* bandwidth of a buck converter is governed by the following relations [4]:

$$8f_{LC} \leq \boldsymbol{BW} \leq \frac{1}{4} f_{sw} \tag{5.1}$$

$$f_{ESR} \leq 2f_{LC} \tag{5.2}$$

where f_{LC} is the resonant frequency of the LC filter for the converter, f_{sw} is the switching frequency of the converter (assuming that it is a *fixed* frequency design), and f_{ESR} is the frequency at which the ESR[1] of the capacitor (and filter resistance) dominates the impedance BW (the bandwidth). All these metrics are determined by the filter and the switching frequency. Often the designer of the power converter will have these data available. However, engineers, though often time constrained, are forced to track down the data for themselves in order to get their jobs done. Additionally, once the inductance and filter capacitance characteristics are known, it is easy to compute these values, as the next example illustrates.

[1] Effective series resistance; see *SRF* in capacitor section in Chapter 4 (Section 4.2.5).

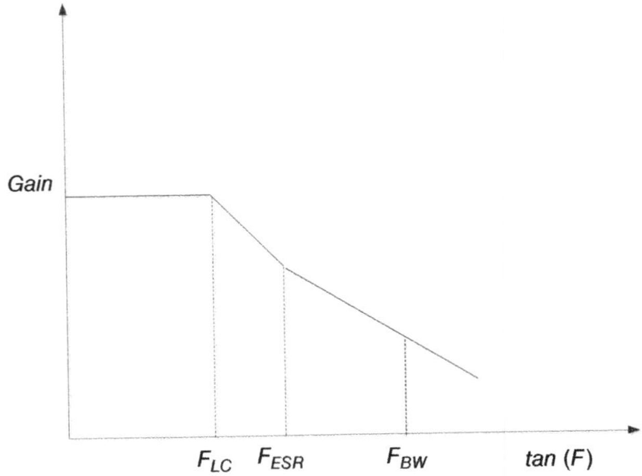

Figure 5.2 Open-loop bandwidth example

Example 5.1 Determine the open-loop response (bandwidth) for a single phase buck converter with a switching frequency of 1 MHz, a storage inductance of 1 μH, and a total output filter capacitance of 400 μF. Assume there are four capacitors, each with an ESR of 10 mΩ. Also assume that there is an additional 10 mΩ of series resistance with the inductor.

SOLUTION: The resonant frequency is determined from the equation for the filter of a buck given in Chapter 2:

$$f_{LC} \cong \frac{1}{2\pi\sqrt{LC}} = \frac{1}{2\pi\sqrt{(1\times 10^{-6})\,(400\times 10^{-6})}} = \frac{10^6}{40\pi} \cong 8\text{ KHz}$$

If the switching frequency is 1 MHz, then the open-loop BW is between

$$63.6\text{ KHZ} \le \boldsymbol{BW} \le 250\text{ KHz}$$

The ESR of the capacitance is ¼ of the total (since there are 4 × 100 μF capacitors in the filter):

$$f_{ESR} \cong \frac{1}{2\pi(R_{ESR}+R)C} = \frac{1}{2\pi(12.5\times 10^{-3})\,(400\times 10^{-6})} = \frac{10^5}{\pi} \cong 32\text{ KHz}$$

which is four times the frequency of the LC filter resonance. Note that the series resistance of the inductor was added here because it shows up in the feedback path and is part of the overall filter. This can be seen in the asymptotic open-loop bode plot of Figure 5.2.

The bandwidth of the converter should not be greater than 250 KHz, which is reasonable given the parameters. Many *current-mode* converters (and some voltage mode) have faster responses due to smaller filters and faster overall closed-loop responses. Their open-loop responses are typically faster due to the smaller filter capacitance that is placed on their outputs and the feedback methods employed [8,9]. For many buck converter designs, this bandwidth is more than sufficient. For power integrity analyses, a basic check of the converter parameters will usually suffice for setting up a time-domain analysis.

Simple Open-Loop Models If the switching behavior of the VR is not critical to the analysis, then a simple L-R-C circuit may suffice for the model. The L-R-C circuit represents the output impedance of the switching regulator. In order to make sure that the converter does not misrepresent the results of the PDN analysis, one must *design* this circuit to behave—at least in the open-loop case—like a converter. Figure 5.3 gives a simple representation of a converter filter. The capacitance C' is represented as a non-ideal capacitor, as in the previous example. This is important because an *ideal* capacitor will distort the final results of the overall time-domain analysis. Some simple capacitor models were developed in Chapter 4 where the non-ideal behavior of a capacitor was modeled. Recall that the series resistance was termed the ESR or *effective series resistance*. The inductance of the capacitor is the *inserted* inductance in the printed circuit board. The storage inductor L, is represented along with the resistance of the interconnect and the "switch" resistance (assuming that FET switches are included in the system). This is what is typically considered to be the R_{ds_on} of the power FET switches—though it is not represented here as a *switching* model in this circuit at this time. Since the two resistors are in series, they can be represented as one value.

The objective is to model the behavior of this structure to allow simulation of the PDN with some level of accuracy. A simple AC analytical model is developed, then, in SPICE to allow examination of the frequency response of this circuit. With the data from Example 5.1, it is an easy matter now to sweep the frequency and determine if the behavior is what is desired.

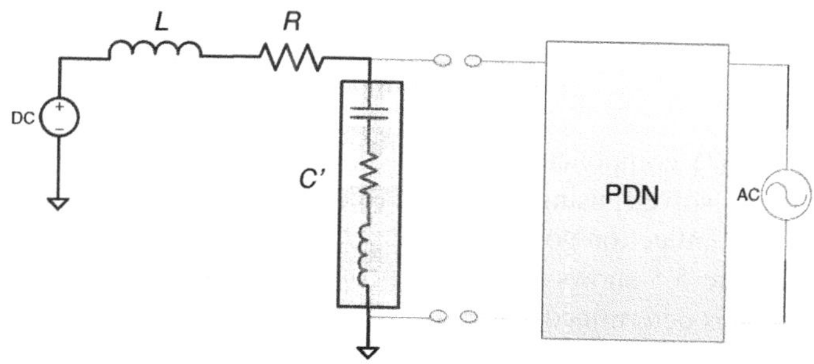

Figure 5.3 Simple LRC model

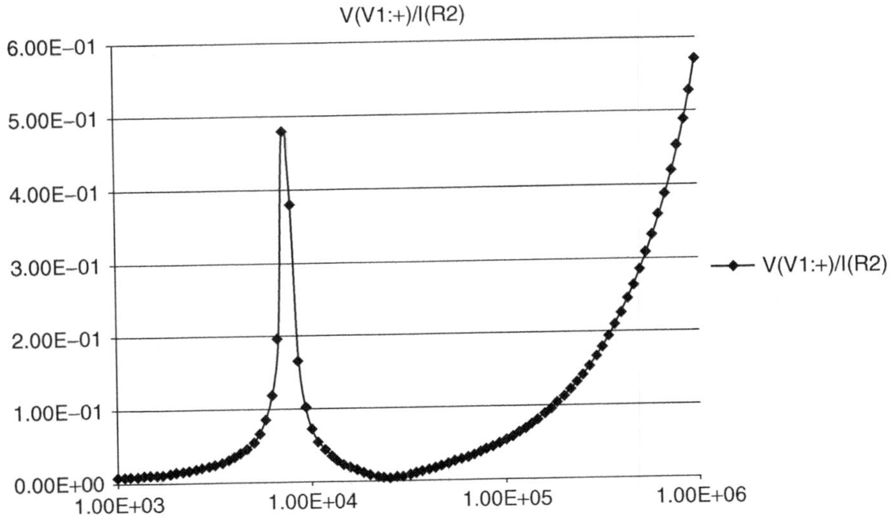

Figure 5.4 Impedance plot of the circuit in Figure 5.3

The setup is shown in Figure 5.4. It is now is simply an AC sweep of the impedance of the circuit looking back toward the voltage source (note that a DC analysis is trivial in this case without a *load source*). The power source is a DC voltage, which may be eliminated from the analysis as the inductor end is grounded because it is an AC sweep (the DC source is effectively an *AC* ground). Notice that the resonance is very close to the simple analysis from the previous example. However, the point of interest here is the *slope* of the curve to the right of 100 KHz. This is where one would expect the impedance to increase, indicating that the VR output impedance is essentially acting like a low-pass filter (delivering a DC voltage) and *blocking* the high-frequency components. Thus, when this model is brought into the PDN model, it will allow one to represent the *filter* aspects of the buck converter correctly. Using the tools from Chapter 4, it is a simple matter to construct the asymptotic bode plot and indicate where the breakpoints occur for this circuit. The impedance function is straightforward and becomes

$$Z(j\omega) = \frac{(R + j\omega L)\,(1 + j\omega R'C - \omega^2 L'C)}{1 + j\omega C(R + R') - \omega^2 C(L + L')} \tag{5.3}$$

where all prime (′) components are related to the capacitor. The roots of the equation can be solved using a simple math program, or even Excel (see Problem 5.1). The inflection points indicate where the changes occur and at which frequencies. Figure 5.5 shows that the impedance increases at the first inflection frequency, which is determined by the root of the first zero in the numerator:

$$\omega = \frac{R}{L} = 1e^4 \rightarrow f = \frac{\omega}{2\pi} \cong 1.6e^3 \tag{5.4}$$

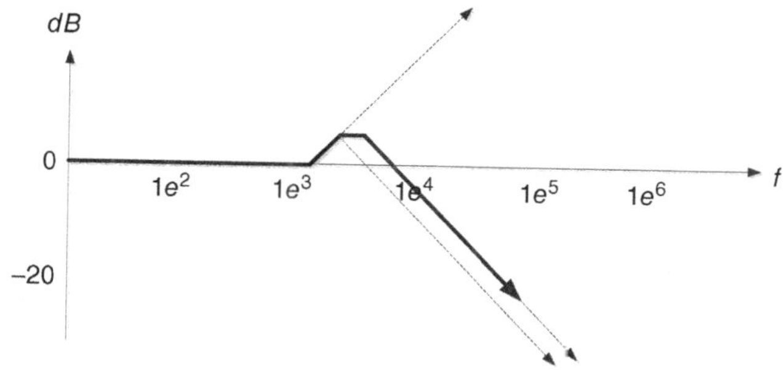

Figure 5.5 Asymptotic Bode diagram of AC

The other frequencies are found using the quadratic formula or any root solver. Note that the actual asymptotic Bode plot shows the *frequency*, not the *angular* frequency ω. Clearly, the simple "ideal" model built as a source for the buck regulator is a *nonswitching* open-loop model.

There are times when it is required to increase the accuracy of the converter representation and therefore build a model that represents switching behavior. We now make a slight change to the model in Figure 5.3 and change the ideal DC source to a *piecewise linear* (PWL or PULSE) function to represent this switching component. The simple SPICE deck for this representation is as follows:

```
* source DISTRIBUTION_1
R_R1 N00248 N00375 .01 TC = 0, 0
L_L1 N00375 N00379 1uH
C_C1 N01254 N00379 400uf IC = 0.96V TC = 0, 0
R_R3 N015181 N01254 2.5m TC = 0, 0
L_L2 0 N015181 100pH
V_V2 N00248 0
+PULSE 0V 2.4V 1us 100ns 100ns 300ns 1us
.TRAN 50us 5ms 2ms SKIPBP
```

This simple open-loop switching model can be used to determine the desirable time-domain results of the PDN once the appropriate stimuli have been added. Here *only* the DC source is changed to a switching source. The switching frequency is the same as that in the previous example. From the *time-domain* simulations for this structure, it can be seen (at steady state) that the output voltage stabilizes around 1 V.

As shown in Figure 5.6, the input waveform is a square wave with the duty cycle defined by the "PULSE" function in the SPICE deck. Note that this duty cycle sets the voltage output as a function of the input voltage. In this case,

$$Duty\ cycle = \frac{V_{Out}}{V_{In}} \rightarrow V_{Out} = (2 \times t_r + PW) \times V_{In} \cong 0.96\ \text{V} \tag{5.5}$$

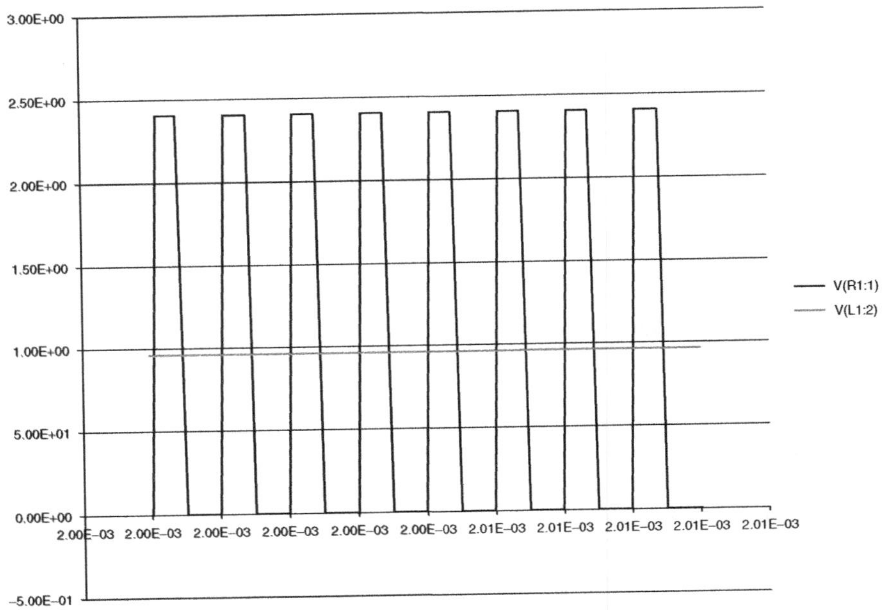

Figure 5.6 Time-domain signals for the simple converter model

where t_r is the rise time of the pulse, and *PW* is the pulse width. The initial conditions were used for the capacitor to enable the simulation to hit steady state a bit faster. A schematic—as compared with a deck—is, of course, the typical way to enter the model, since all the tools today have easy-to-use GUIs.[2] The SPICE deck here was an export of the schematic from PSPICE, which is identical to Figure 5.3, except for the PULSE function call. The next section examines some simple closed-loop systems, expanding upon the discussion on open-loop models.

Simple Closed-Loop Model for the Buck Converter There are a number of different ways to represent a buck converter with a closed-loop feedback system [5–7]. The closed-loop model is a *nonswitching* model, using a simple *voltage-controlled voltage source* with a feedback amplifier, shown in Figure 5.7. The amplifier is connected to the load voltage (typically the silicon-device source), which is the controlled source for the closed-loop gain stage element ***E***.

When the load changes, and voltage across the amplifier also changes, the voltage source ***E*** then increases its voltage to maintain the output at the right value—as in the controller for a simple buck converter discussed in Chapter 2. Both the voltage-controlled voltage source and the amplifier are considered *ideal* for this analysis. Again, as with open-loop systems, the objective is to represent the converter in such a way that it does not disrupt the behavior of the PDN and overall

[2]Graphical User Interface.

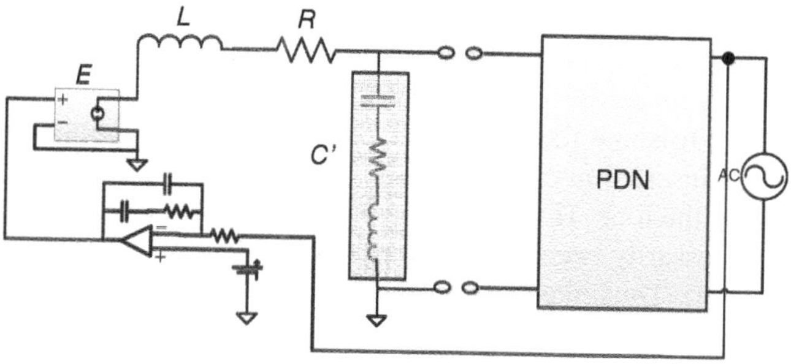

Figure 5.7 Closed-loop simple feedback model

system in the time-domain. The first step is to determine the gain of the voltage-controlled voltage source. As before, the addition of the PDN will be neglected, in order to develop the model of the converter, and added later. As in Chapter 2, if the intention were to develop an actual closed-loop switching model, a triangular wave generator and PWM circuit would be inserted into the feedback loop in addition to the compensation stage. However, this has been eliminated (in this analysis) for simplicity. However, it is still required to set the overall gain of the voltage-controlled voltage source. The gain is found from the simple formula below,

$$E_{Gain} = \frac{V_{In}}{V_{Tri}} \tag{5.6}$$

where E_{Gain} is the controlled source gain (the goal here), V_{in} is the input voltage for the converter, and V_{Tri} is the peak-to-peak voltage for the triangular wave generator. Since this triangular wave generator is not in the circuit (yet), an arbitrary value is assigned to it. As a practical concern, it is not desirable to have a gain that is excessive, since this could shift out the compensation point to where it could be difficult to compensate the feedback. Thus, it is good to choose a value for the triangular wave voltage to be representative of a real circuit, say 1 V. The input voltage would typically be the input voltage for the converter. Here, a value that is part of the design is assigned. For this example, the same value will be used as before (e.g., 2.4 V).[3]

Now that the gain has been set, the loop may be compensated, which was the initial goal of this analysis. The first step here is to create the Bode plot for the system. The desire is to compensate the feedback amplifier so the converter operates in closed-loop. Note that in Figure 5.7 there is a *type II* compensation for the feedback represented. One could have just as easily placed a *type I*, or a *type III* compensator here. The choice is dependent on what type of response is desired for

[3]Most platform converters use 12 V as their input voltage. Many smaller consumer electronic platforms utilize lower voltages for their inputs, which then change the duty cycle and thus the compensation network. The unusual voltage of 2.4 V in these examples shows that the methods are the same, regardless of the voltage ratios.

the system and the effective bandwidth of the filter. Since the intention here is not to *design* the converter feedback circuitry explicitly, but to represent it correctly for the power integrity analysis, the closed-loop bandwidth represention for the converter is sufficient to show that the loop response mimics the converter correctly when the PDN is inserted into the analysis. There are a number of ways that one could compensate the loop. However, the goal is to achieve a reasonable representation and not necessarily become a converter design expert (that will be left to the power design team). Thus, the use of *Veneable's K-factor method* will be applied to compensate the loop [10].

The gain/phase plot is shown in Figure 5.8. Instead of showing the SPICE deck again here, a screen capture is taken to show the schematic for the setup in Figure 5.9—this is the actual circuit that will be modeled in PSPICE. The unity gain crossover point is around 14.4 KHz, and the phase margin at this point is 161 degrees. (Note the similarity in the magnitude of the bode plot with respect to the asymptotic version.) However, the interest here is in finding the open-loop gain at the projected operating bandwidth. This is typically determined from the frequency of the RC network.

The bandwidth in this case was determined from the computation in Example 5.1, which yielded 32 KHz. One could stabilize the loop below the resonance of the filter (~8 KHz), which would guarantee stability, since at this point the gain is already in the system. However, the bandwidth is desired to be a bit higher (in response to the load changes), and thus one can compensate the filter above this resonance. For this reason, a *type III* compensation is chosen, which will raise the expected closed-loop bandwidth to be more in line with the expected closed-loop response of the system—something that the power converter designer would be expected to achieve. One could simply ask the power design engineer to give the

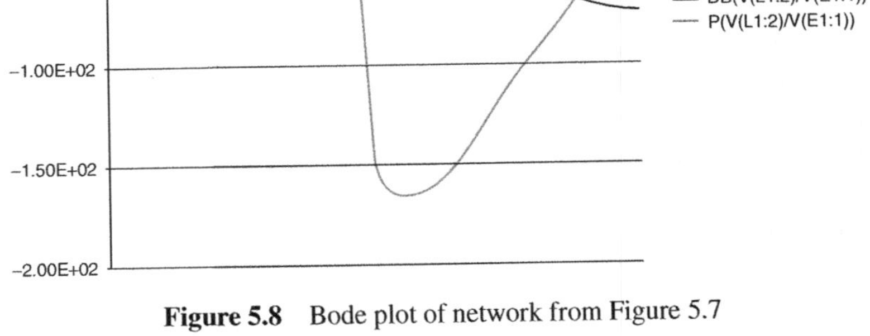

Figure 5.8 Bode plot of network from Figure 5.7

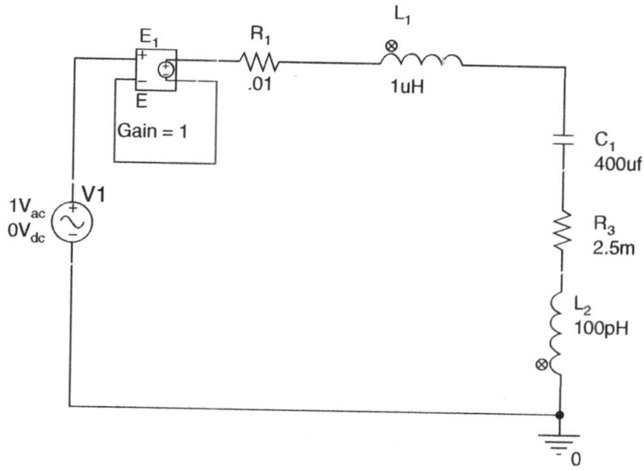

Figure 5.9 Schematic for SPICE setup

closed-loop bandwidth of the converter—however, as stated previously, often the data are not at the fingertips of the design engineer.

Examining the Bode plot once again, the goal is to determine the cross-over frequency for where the desired closed-loop gain needs to be at unity gain (0 dB). This is the point at which the loop will be closed. Given that it is to be closed with a higher bandwidth than the resonance, gain will need to be added to the system. The choice for the closed-loop *BW* is somewhat arbitrary. From Example 5.1, the bandwidth of the system was desired to be below 32 KHz. From the Bode plot, it is clear that the phase starts to rise around this point. Thus, if the loop is closed at 30 KHz, the compensator should be able to respond adequately.

The next item in this procedure is to determine the *phase margin*. In most literature, 45 degrees of a phase margin is adequate (many real converters opt for a higher margin to increase the range of stability, but this is a good typical value and is adequate for this analysis). At 30 KHz, the gain has dropped by about −15 dB. Thus, the amount of gain that will be needed to add with the error amplifier will be

$$Gain = 15 \text{ dB} = 20 \log(A) \rightarrow A = 10^{15/20} \cong 5.62 \tag{5.7}$$

At 30 KHz, the *phase boost* (PB) of the system is now determined, which will help give the *K-factor* that will be used in the equations:

$$PB = PM - PS - 90 = 116° \tag{5.8}$$

where all the values are in *degrees* rather than in *radians*. *PM* is the *phase margin* (45 in our case), and *PS* is the open-loop *phase shift* at the frequency where the loop is closed (∼161°). Figure 5.7 now shows the actual error amp section that will be used for the analysis, with results shown in Figure 5.10. The feedback amplifier

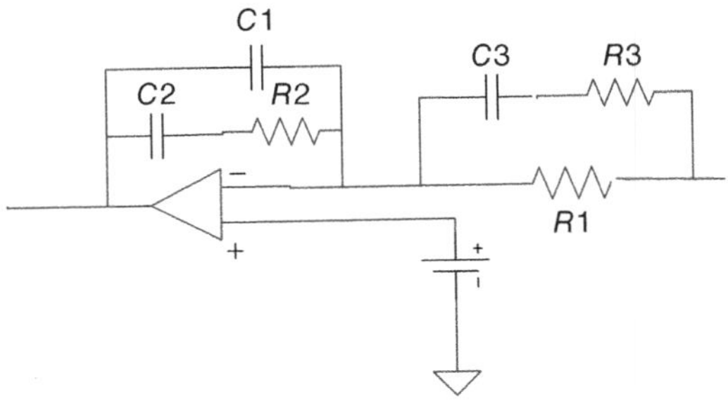

Figure 5.10 Error Amp from Figure 5.7

has six components that need to be computed; R_1, R_2, R_3, C_1, C_2, and C_3. The values (including the *K-factor*) are computed from the formulas given below [10]:

$$K = \left\{\tan\left(\left(\frac{PB}{4}\right) - 45\right)\right\}^2$$

$$R_1 = 10\ K\Omega,\ R_2 = \frac{\sqrt{K}}{2\pi f C_2},\quad R_3 = \frac{R_1}{K-1} \tag{5.9}$$

$$C_1 = \frac{1}{2\pi f G K R_1},\quad C_2 = C_1\ (K-1),\quad C_3 = \frac{1}{2\pi\sqrt{K} G R_3} \tag{5.10}$$

The values may easily be determined from a spreadsheet or a Mathcad program (see Problem 5.2).[4]

Now that the design is done, the values for the circuits are put into the SPICE model. Figure 5.11 shows the circuit as it is represented schematically in SPICE. In the figure, the loop is *broken* to determine the open-loop gain and phase of the system. The Bode plot in Figures 5.12–5.13 shows both the phase and gain plots for the open loop, closed loop, and complete loop (additive) in the system. Note that the phase margin at 30 KHz is

$$PM = 180 - 135 = 45° \tag{5.11}$$

This was the design point for the compensation network.[5] In general, the design of the compensation loop is dependent on a number of factors, but only basic tools were used to develop one here. A more thorough development (as one would expect) would be done by the power engineer who is actually *designing* a converter. A closed-loop switching model is now developed using the design tools that have been applied thus far.

[4]A type II compensator could be used for this design; however, the bandwidth then should be reduced. Reviewing the design equations is a good exercise, as it can show where the iterative approach is necessary.
[5]Note that because the ESL of capacitor is not added for computation for Figure 5.13 and thus, both a phase and gain difference will exist between this figure and Figure 5.8.

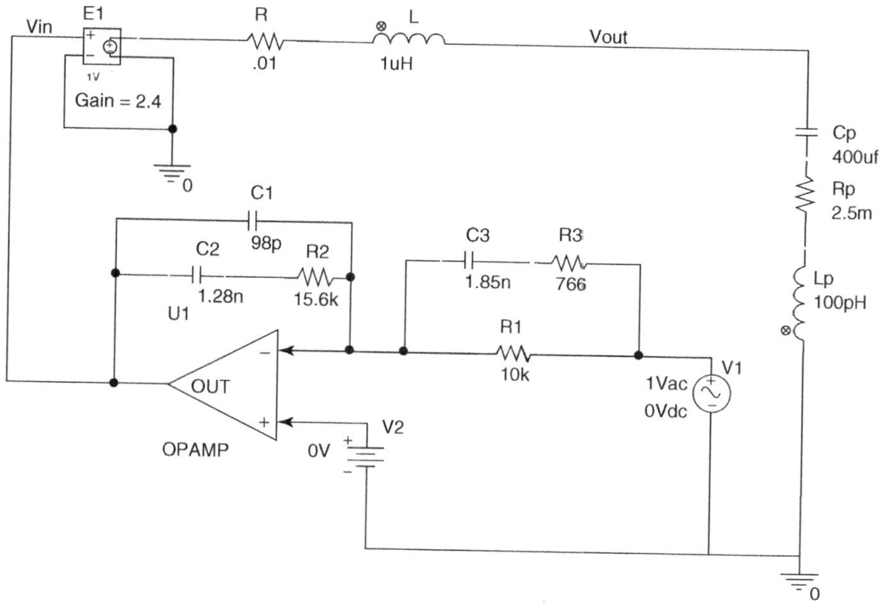

Figure 5.11 SPICE level schematic of closed-loop nonswitching model

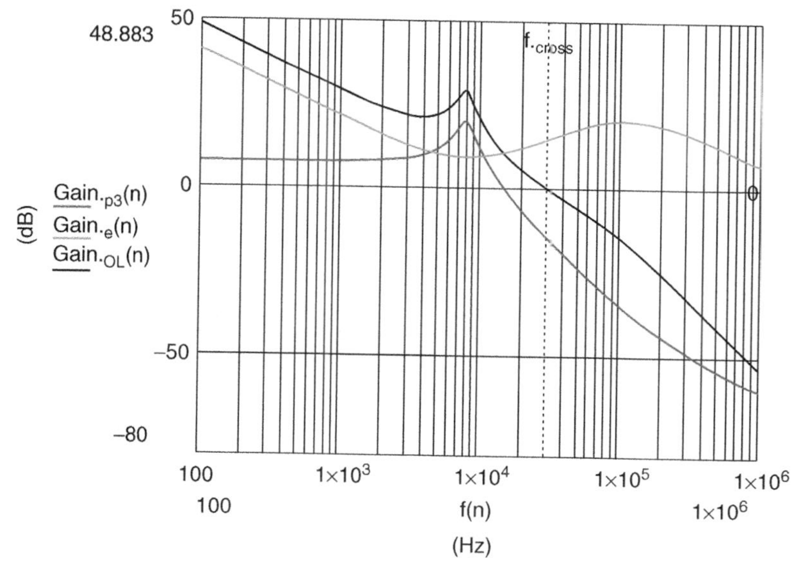

Figure 5.12 Gain plot of closed-loop network for Figure 5.11

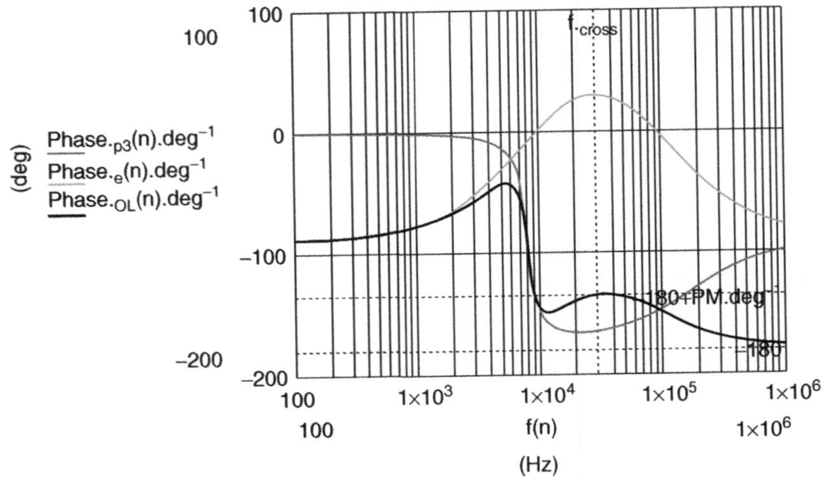

Figure 5.13 Phase plot of closed-loop model from Figure 5.11

The closed-loop switching model is simply an extension of the nonswitching model. The closed-loop Bode plots should in fact be the same given that the only change is the switching components added to the model (this is the typical way to determine the loop response of the system, since adding the switching components makes it impossible to perform an AC sweep). The three key circuit components that are added, are the triangular wave source, the PWM comparator (which converts the triangular wave to a PWM signal), and the switches.

A first-level schematic is shown in Figure 5.14. The triangular wave generator is a simple PULSE waveform that feeds into an ideal comparator circuit, which

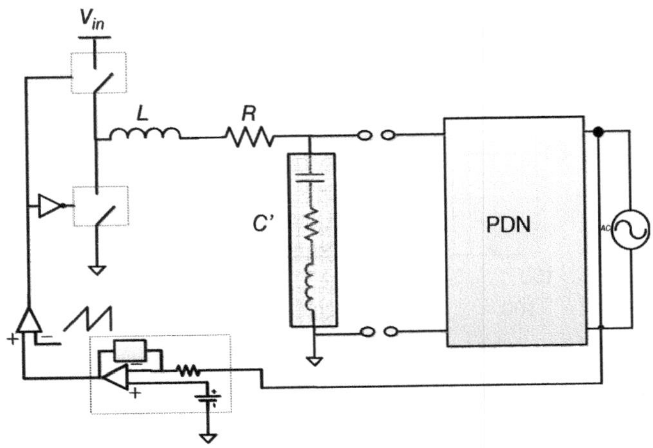

Figure 5.14 Full closed-loop simple model representation

creates the PWM square-wave pulse. The compensator is the same as before and feeds into the other side of the comparator. There are a number of ways to generate an inverting signal for one of the bridges. Here, it is shown as an inverter to the input signal. In reality, another comparator circuit would have been used.[6]

The complete SPICE model, along with some key waveforms, is shown in Figure 5.15, with the parameters displayed for the key elements. There are a number of ways to construct this closed-loop model, but it is beyond the scope of this book to illustrate them. The reader is encouraged to review the references at the end of the chapter for a thorough discussion of the different methods to model the closed-loop behavior.

The important thing about this model is that it correctly responds to a dynamic change in the output. Figure 5.16 shows the PWM signal at the node (V_x), the triangular wave, and the output voltage (V_o). Note the *diode* conduction effects at the switching edges of the *ideal* switches. There are a number of SPICE related anomalies that are added to the model to allow it to converge (as shown in Figure 5.15). Details on these devices can be found in the references listed at the end of this chapter. However, the reader should be careful of having the gain of the amplifiers become too large (e.g., exceeding 10 K), as this can cause convergence issues. The reader should also be aware of convergence issues in the time-domain with the use of inductors. These common problems are easily addressed if one heeds the warnings. We will return to these models in a later section where some additional time-domain system analyses are discussed.

5.1.2 Load Representations

Now that some basic buck converter source structures have been examined for source modeling purposes, it is appropriate to look at different load constructs to round out the time-domain portion of the analysis. The load analyses for time-domain simulations are representations of the *time-rate-of-change* of a load during *worst case* activity in the silicon. Every device exhibits different load behavior, and thus the analysis must start by determining what the device *load* characteristics are before the PI engineer can start the time-domain simulations. Therefore, it is best to begin by examining what *causes* a load change to occur in the first place.

Load Behavior Usually, the largest dynamic loads in the industry are those generated by microprocessor cores. Figure 5.17 illustrates a simple block diagram of a generic microprocessor. For most processor cores, the most dynamic portion of the load—with the greatest dynamic load range—is the execution unit of the core. This is where the bulk of the work is typically done for many workloads [11].

Figure 5.18 shows a topographical map of a processor core where the dominant current loading would occur. For most architectures, the load is dominated by the execution region during the processing. Another view in Figure 5.19 shows a

[6]Note that this is just for single-phase operation. As in Chapter 2, separate controls would be used for a multi-phase solution to ensure that the duty cycles to each bridge was correct.

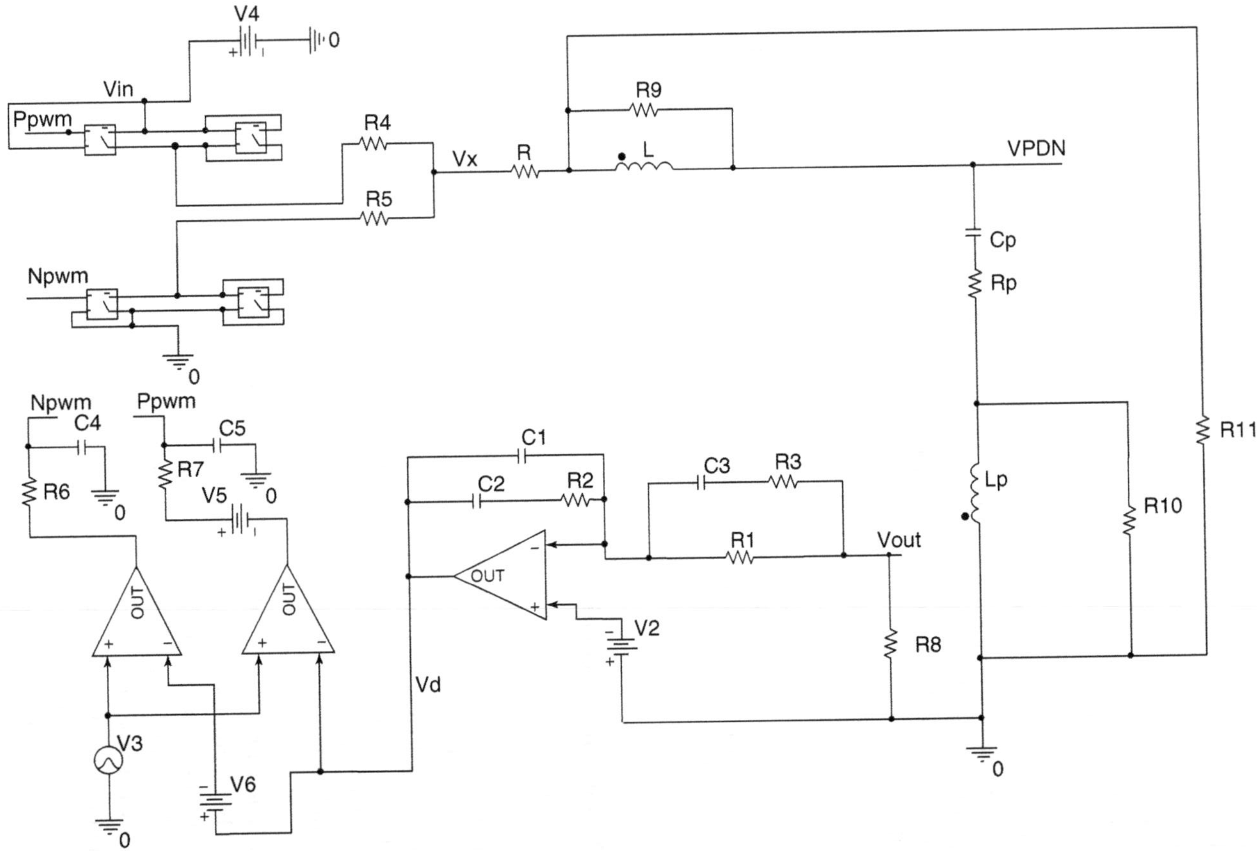

Figure 5.15 PSPICE schematic of closed-loop single-phase buck regulator

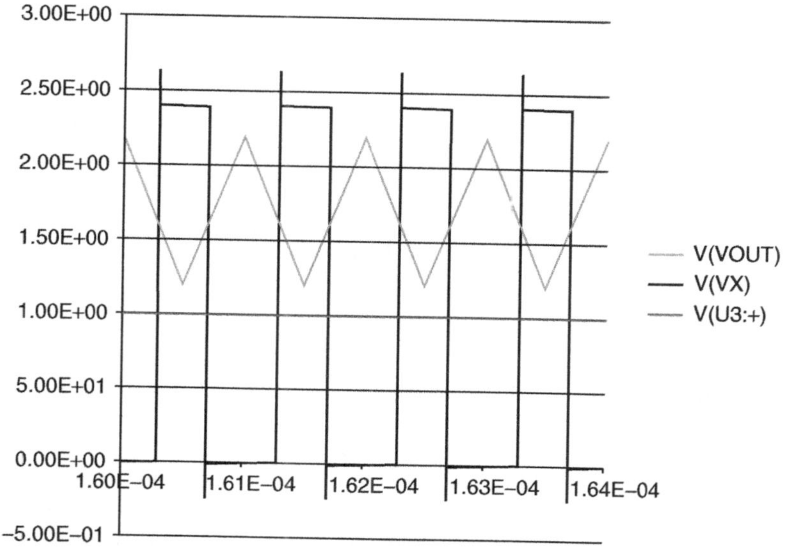

Figure 5.16 Selected time-domain waveforms for model in Figure 5.15

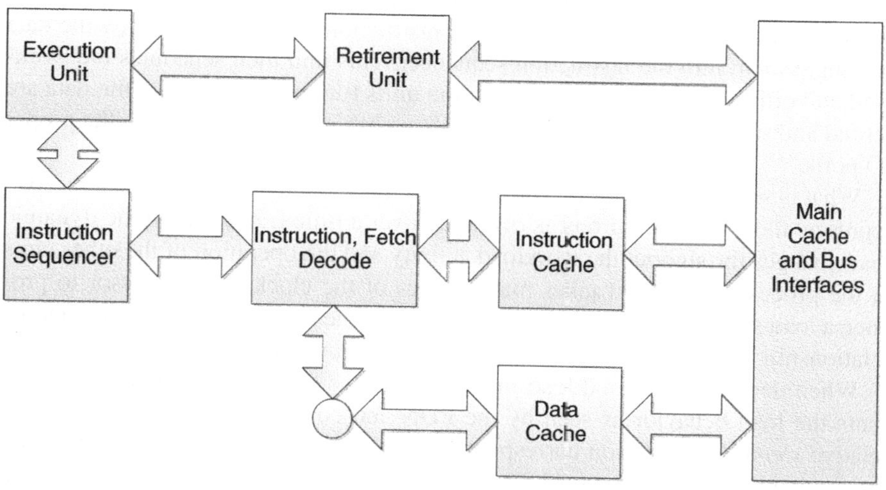

Figure 5.17 Simple generic microprocessor core block diagram

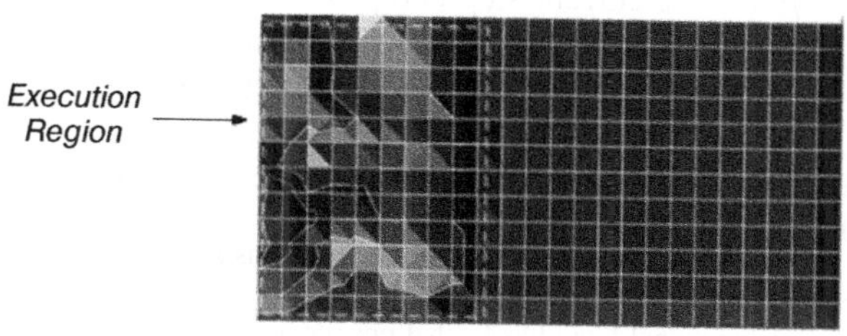

Figure 5.18 Topographical map of processor loading

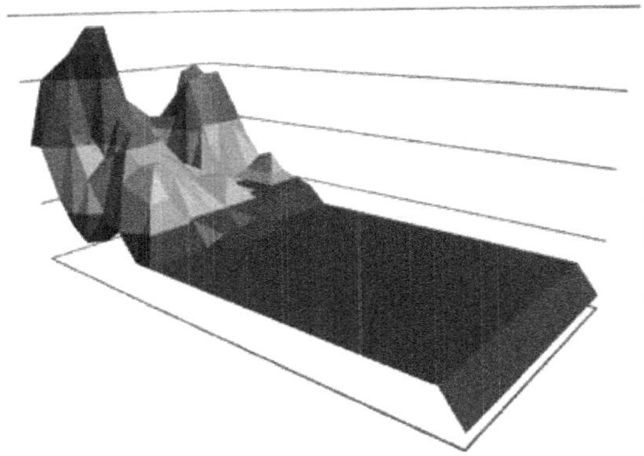

Figure 5.19 Topographical current load profile of microprocessor core

three-dimensional topographical load profile of a processor core. It is instructive to examine this topographical current load profile because it also illustrates the corresponding load behavior as seen by the power distribution network. The large spikes are where the execution unit is generally located.[7]

Depending on the workload, processing occurs in one or more of the execution blocks (integer, floating point, etc.). The instruction and fetch unit *fetch* the data and input them into the instruction sequencer. This unit then schedules the workload and orders it into the proper execution units for processing. Last, the data are retired and sent to the memory and data busses, which direct the data to the proper IO ports.

What is seen by the PDN is typically a filtered version of the instantaneous load representation, and the current is *averaged* over a time frame due to the dynamic load changes throughout the workload activity and the operation of the other units in the processor. Often, it takes many cycles of the clock in a processor to produce a load step change that manifests itself onto the board power network, and its relationship to the actual execution load is usually indirect.

When developing a good load model, the key is to create a profile that represents the load behavior as seen by the PDN and converter. Figure 5.20 shows a relative *stepladder* function corresponding to the core clock that is running in an *execution* block of a processor.[8] The clock and data throughout the *cluster*[9] logic make stepwise increases, which result in a load profile that mimics an overall step load change *I_step*. As the current step is repeated–usually in a nondeterministic fashion—over a given period of time, the local behavior becomes *harmonic* (this load behavior will be discussed more in a later section). The important thing to

[7]Figure 5.19 does not show an actual core—it only shows where loading may occur on a device. The legend refers to only relative loading behavior.

[8]This is a generic definition, since the aggregate load behavior as seen by the PDN would be nearly impossible to describe.

[9]The cluster consists of functional logic blocks that operate as a group. For example, the execution cluster may comprise a floating point and integer block along with other interfacing logic.

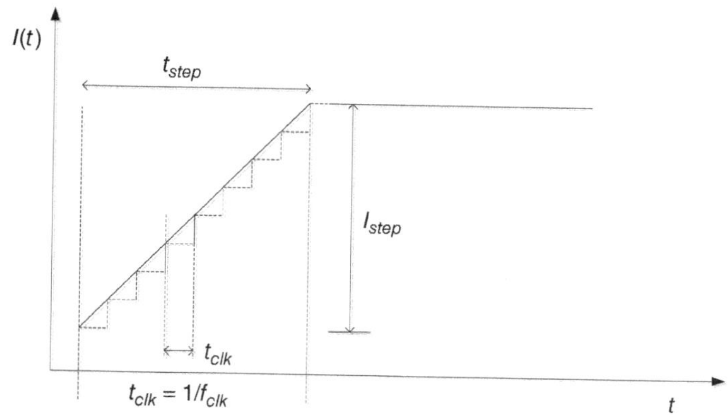

Figure 5.20 Current step in relation to core clock

remember is that this gives us a fair representation of what the actual load will entail.

The actual load step profile (as similar in Figure 5.20) is found through a combination of empirical data, measurements, and estimates by the silicon team in charge of that device. (Often design teams create modules, that mimic the load behavior of the actual silicon within the system. These devices are placed into *socketable* locations where the silicon is attached—or soldered down to test boards when sockets are not part of the design—so that PI engineers and others can examine the PDN and system responses to dynamic loads [14]). The result is a load profile that represents a worst-case load step and slew-rate over a given time to allow some margin in the predictive process by the team. The estimate always has some *guardband* so that the PI engineer can safely develop the network to support the power distribution into the silicon load. Unfortunately, if the margin is too large, it may result in too *stringent* a PDN design, which can cause problems at the converter, the PDN filtering, or both. It is therefore always a good idea to get these data (load information) initially from the silicon team and then iterate the profile through the design and analysis process.

From a silicon designer's perspective, it is critical that the power delivered to the load be sufficient to power the device under all normal operating conditions.[10] The two most important characteristics of this are the *slew-rate* and maximum *current step*. The profile in Figure 5.20 is defined by the parameters in Table 5.1. Most processors are *multi-core*, and thus, it is not just one core or processing unit that the PDN sees—it is *all* of the logic level load placed on that power domain.

[10]There are some operating conditions, such as a virus, that are considered abnormal. These conditions are typically associated with errors in the system and can cause a shutdown of the system. However, the PDN and voltage source are typically designed to handle such conditions and bring the processor down elegantly.

TABLE 5.1 Current load profile parameters

Parameter	Definition
Δt_{step}	Time length of current step
I_{step}	Maximum current step
I_{smin}	Minimum current for I_{step}
I_{smax}	Maximum current for I_{step}
T_{per}	Period of current waveform
$I_{\text{step}}/\Delta t_{\text{step}}$	Slew-rate of step

The silicon power designer has to only give the slew-rate and the maximum step, and let the PI engineer design the PDN and the system to meet this load profile. However, it is best to define all of the parameters to allow the PI engineer to properly model the entire power path. The length of the time-step, Δt_{step}, is considered the *width* of the rise time (0 to 100% in this case) and is usually defined from an operating current load I_{smin}, to the highest, I_{smax}. These are *not* the minimum and maximum current operating points; they are only the two points in the current step that define the worst-case step function. This is an important distinction because it is easy to confuse these data with the actual operating range of the silicon device. The processor will typically operate over a much wider dynamic range due to different operating modes of the processor—active and inactive. The periodicity of the current load pulse is determined by T_{per}. Usually, this is a generically defined value found through empirical data from the previous generation of devices and based on some of the more recent estimates found on the current generation of a processor design. This periodic time-domain profile can be very complex and is highly dependent on the activity (at least indirectly) of the processor workloads and full processor system operation. It is often *not* defined by the power silicon designer because it is very difficult to construct given that it is workload dependent. This profile is discussed in a bit more detail in a later section. The slew-rate, as shown in the figure, is simply the ratio of the current step to the time step. Either of these two values is determined by the other once the slew-rate is defined. The important thing to note here is that the current step must be completely defined to allow the PI engineer to start the process of building the model and getting results.

Load Models Now that the load behavior has been examined (at a high level), it is possible to start to put together an estimate for the load model. A number of different models in SPICE are available. Figure 5.21 shows four that are frequently used for modeling a current step or series of pulses. It is beyond the scope of this book to go through every possible load model that could be used here, and the reader is encouraged to review the SPICE references [1–3] to gain a better understanding of the capabilities of each. Moreover, the objective here is to show only a few of the models that are available. For current load modeling, the use of the G model will be illustrated, which allows one to model the load as a *transconductance* source. It

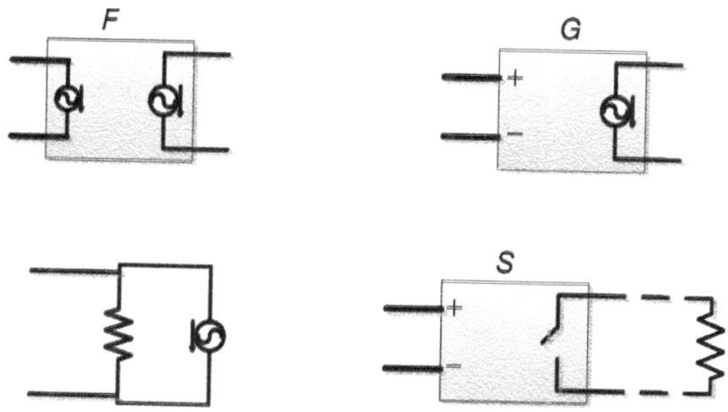

Figure 5.21 Different model sources for load representations: Starting from top left—current-controlled current source (F), voltage-controlled current source (transconductance G), simple current source with resistor, and voltage-controlled switch (S)

could have been just as easy to start with a simple current source and then iterate to a different model if that suited the analysis.

For more sophisticated analyses, the PI engineer could also use a polynomial (e.g., an *Epoly* model) to model the current source. This may more accurately represent the wave-shape of the current source—provided that the information is available. However, for most cases, the basic model of choice is somewhat arbitrary, so long as it generally represents the load's behavior. The transconductance model allows one to control the load current as a function of the voltage deviations. For the ***G*** model, the relationship between the current and voltage is,

$$I_S = \gamma V_C \tag{5.12}$$

where γ is the conductance (inverse of resistance) in Siemens and V_C is the controlling voltage source. Thus, if one puts 1 V across the controlling source with a transconductance of 4 Siemens, a total current of 4 amps will be put on the source. The G model typically converges (as will be seen in the next section) reasonably well under most conditions, and the control source is easily implemented with a PWL function or PULSE voltage function.

The values or metrics used to define the wave shape (generically) are those listed in Table 5.1. This type of current step model, though more complex than a simple resistor, will yield a reasonable step response for the system—and this is where one typically begins when starting a system PDN analysis. Once the step response is defined, and the system PDN analysis begins some level of iteration will result between the platform design engineers recommendations and the silicon design teams which ends up in one or more of them trading off various parameters.

Figure 5.22 shows a test circuit for the G model. The controlling source is a PULSE function with a few parasitics added to show the effects of the step function, the results of which are found in Figure 5.23. The step function current source magnitude is 1 amps and the slew-rate is 1 A/100 ns. As expected, with a small filter, a droop event occurs, which is very evident in the figure. As the amount of

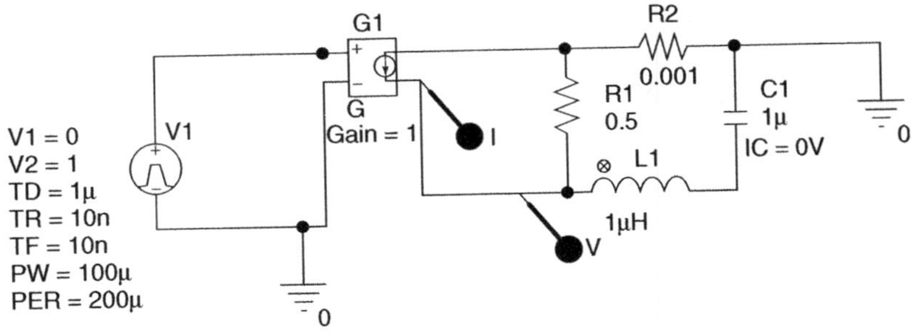

Figure 5.22 Illustrating use of G model for current load

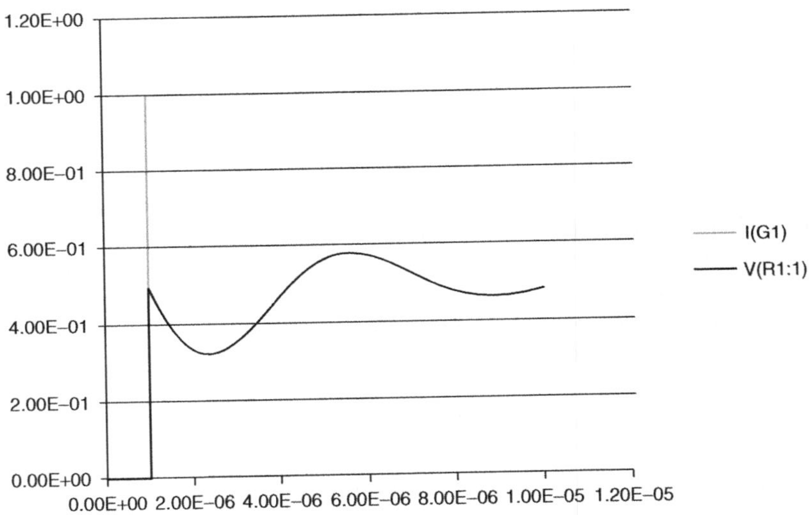

Figure 5.23 Transient response from G model test circuit

the filter (increase in capacitance) changes, the droop can be expected to decrease. This effect was examined in the discussion of the frequency-domain in Chapter 4. When one adds additional filter components e.g. mid frequency, low frequency, etc. this has an effect on the distribution resonance peaks and thus will have an effect in the time-domain results. These time-domain effects will be examined more in the next section.

5.2 TIME-DEPENDENT SYSTEMS

Time-domain analysis has traditionally been the main task of power integrity engineers and has often been considered the *only* deliverable by the PI engineer for many projects. As has been seen, this is just one of many analyses that a PI engineer will accomplish, but still an important one.

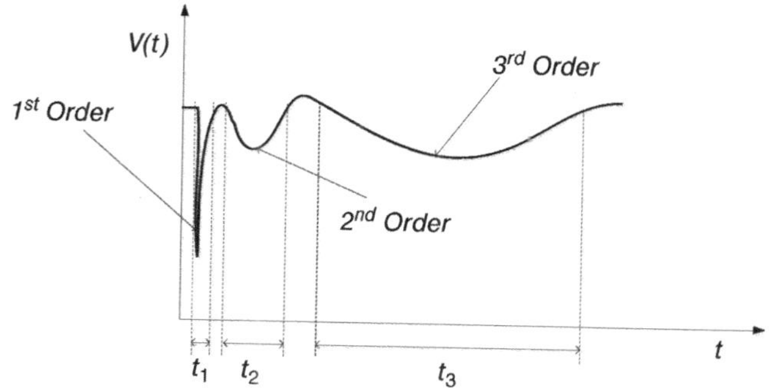

Figure 5.24 Voltage droop definitions

5.2.1 Voltage Bus Droop Boundary Conditions

The time-domain analysis of the PDN starts with understanding the boundary conditions of the voltage droop in the bus. This is essentially a check on the efficacy of the entire PDN and power source—as defined and/or designed [12]. Figure 5.24 shows the basic definition, as seen at the load, of the power distribution network *droop*. In a real system, one would try to measure this droop at the point nearest the silicon load—or as practically as close as could be achieved. Usually, this means measuring the droop between the power and ground pins of the silicon device [13].

The voltage droop is decomposed into the frequency domain from which there can be inferred which droop elements and which portion of the current step are dominating. This was seen to some extent in Chapter 4, where the PDN was analyzed in the frequency domain with the intention of observing which elements contributed to which portion of the impedance profile the most. For the definitions in Figure 5.24, it is a straightforward exercise to examine which portions of the curve are dominated by the elements in the PDN and load. The first droop is typically driven by the silicon-level capacitance, the local resistive parasitics on the silicon, and the silicon-to-package connections. Typically, this droop has a duration of 10s of nanoseconds or less. The second-order droop is dominated by the mid-level capacitance on the package and interconnect of the package planes—and sometimes the connector pins. It is typically anywhere from 100 nanoseconds to one microsecond. The interesting part of this droop event is that the pulse frequency is effected by many of the filter aspects on the package and the silicon. The third-order droop is usually within the bandwidth of the converter (or close to it) and is 10s of microseconds or longer. The converter capacitance, as well as some of the bulk PDN capacitance and interconnect, dominate this. The third-order droop is usually slow enough that the bandwidth of the converter may respond to it. For the bulk of the converters in computer platform designs—and mainly due to the PDN—both second- and first-order droops are outside of the bandwidth of these voltage regulators, and thus it is up to the filter response to squelch the amplitudes of these droops.

Figure 5.25 provides a sketch of an impedance profile, illustrating where the impacts on the droops reside. The resonances do not perfectly correlate with the

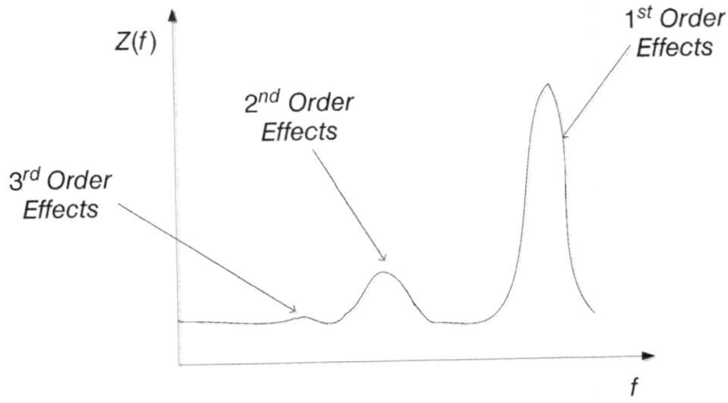

Figure 5.25 Example impedance profile and droop event relationships

TABLE 5.2 Example of relative droop percentages

Droop	Percentage of V_{nom}
First order	0.15
Second order	0.12
Third order	0.10

droop values but the *range* of frequencies still does. The rationale for limiting the droop events will become evident in the next chapter. As was seen in Chapter 4, each element in the distribution network affects the overall impedance, and thus, the particular way in which the PDN droop responds. The objective here is to understand how to design and analyze a PDN that results in a droop profile that falls within the specified V_{max} values set for the silicon. These values depend on many factors, including the operational range of the process that the silicon is based on. The silicon design team may choose to pre-set these values and force a duration on each droop associated with it. Table 5.2 shows some typical values as a percentage of the nominal voltage associated with the droop.

The tolerance to the droop events is dependent on both the activity of the event and the duration. Typically, first-order droop events are less than 20 nanoseconds (as stated above), and thus, the silicon device can withstand a larger droop during this shorter timeframe. This is possible because many devices can adapt to short duration droop events by *slowing* the frequency of the device, which insures that data integrity is preserved. As the droop events lengthen and get larger, it becomes much more difficult to maintain data integrity, and thus, the need for different definitions and limits to each droop event [15] must be enforced.

5.2.2 Voltage Bus Droop Boundary Analysis

From the previous information, it is possible to now put together the full system for a PDN time-domain analysis. The data and methods from the previous sections will be used to build a system model. Since there is already a representative PDN from Chapter 4, it is a simple matter to build a similar one and use an appropriate source and load model to perform a droop analysis. The next example illustrates this process.

Example 5.2 Determine the droop using the PDN from Section 4.3.2 in Chapter 4 and the open-loop source model at the beginning of this chapter. Assume that the output voltage is 1.2 V at the load. Use the G model and the slew-rate data in Table 5.3 for your analysis. What are the droop values for the PDN? Do they meet the specifications in Table 5.2? If the VR capacitance is increased by 5*X*, what happens to the third-order droop? Would the closed-loop VR designed earlier in Section 5.1.1 work (e.g., respond fast enough to mitigate the third-order droop event)?

SOLUTION: The schematic for this is left as an exercise at the end of this chapter. In the first view, shown in Figure 5.26, the third-order droop is clearly a large event, and that if this pulse is treated as a half *sine-wave*, the frequency of this pulse will be approximately in the 6 KHz frequency range. However, note where the droop event starts to increase—beyond 120 millivolts (the third-order droop voltage limit). This occurs faster than the bandwidth of converter (30 KHz), and thus the VR needs to respond much faster to this droop, or one will need to increase the filter capacitance, or both. Based on the previous analysis in section 5.1.1, it is clear that the VR will not be able to respond to this event without some changes.

The next view, in Figure 5.27, shows the second-order and first-order droop events. Note that these droops occur over very short durations—as would be expected from the fast edge rate. Also, note that all the droops are *referenced* from ground rather than from the mean voltage. For this simulation, it is easier to simply use a 0 V source, since there is no need to wait for any settling due to the source charging the capacitors in the system (one could also use the initial conditions on the caps, but these need to be known *a prior*).

At this point, there are a couple of options for the PI engineer. The first is to increase the decoupling capacitance throughout the PDN. The other is to simply

TABLE 5.3 Step-function values for simulation in Example 5.2

Parameter	Metric
Δt_{step}	2 ns
I_{step}	20 Amps
I_{smin}	0 Amps
I_{smax}	20 Amps
T_{per}	N/A
$I_{step}/\Delta t_{step}$	10 A/ns

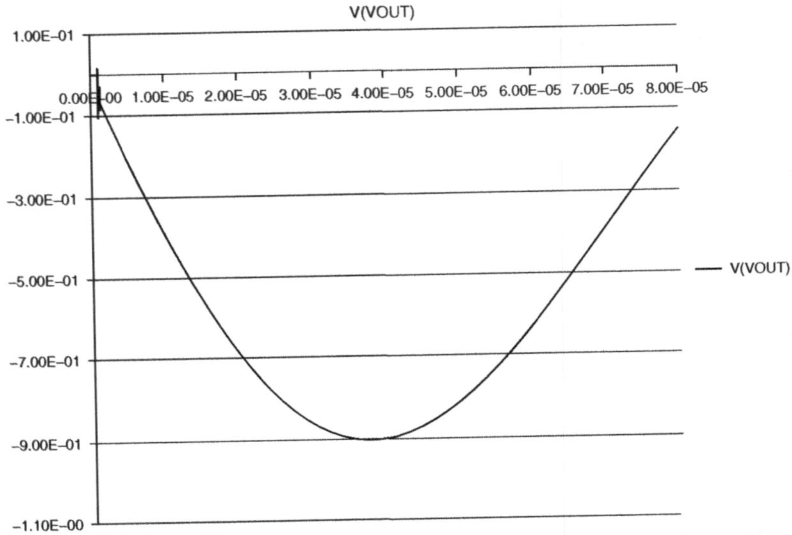

Figure 5.26 Third-order droop event

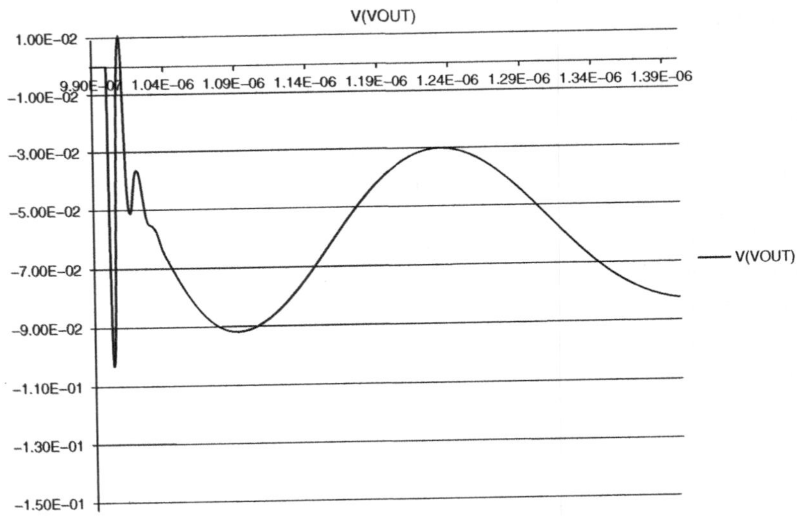

Figure 5.27 First- and second-order droop events

increase the converter capacitance, as the directions stated. This means that the compensation design for the converter must be re-done, but not for the open-loop case, where the effect can be seen and it is easy to determine if that does the job quickly. It is clear from Figure 5.28, that increasing the VR capacitance had a large effect. However, this is still not enough to mitigate the droop if the bandwidth of

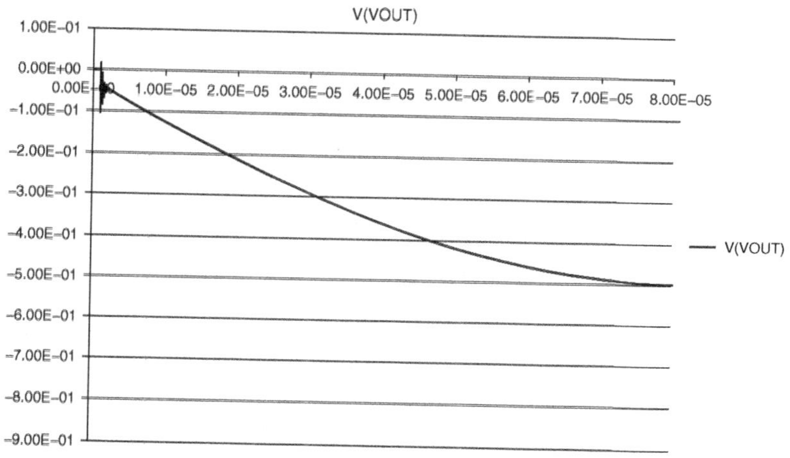

Figure 5.28 Third-order droop event after increasing VR capacitance

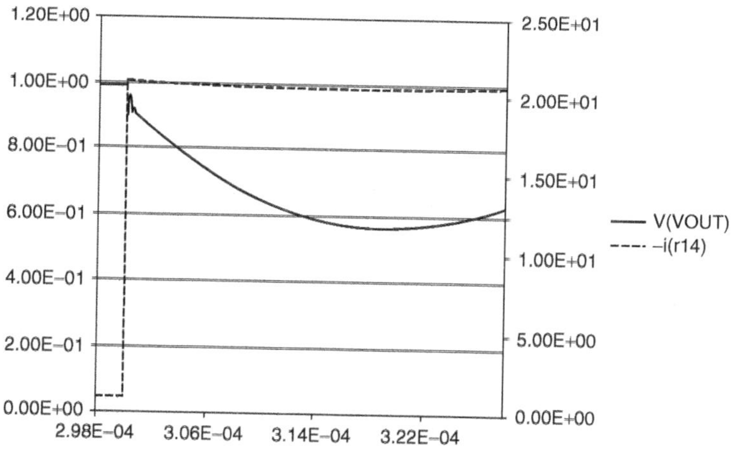

Figure 5.29 Third-order droop with current step on CL switching model

the converter is only 30 KHz. Thus, a combination of additional capacitance to the PDN and additional filter capacitance for the VR is necessary.

Finally, for completeness, the same PDN is shown with the *closed-loop* model in Section 5.1.1. As can be seen in Figure 5.29, the response to the droop event comes after approximately 20 microseconds. However, by this time, the voltage has dropped by nearly 500 millivolts! The good news is that with the addition of the filter capacitance in the PDN and converter filter, this droop can be mitigated. This is left as an exercise at the end of the chapter.

The PDN time-domain analysis in Example 5.2 showed that an iterative process can be used to meet the complete design criteria of a time-dependent system. Sometimes, the modeling effort calls for relaxing specifications with the silicon team. It may even require going into a much more sophisticated model than the one presented here. The simple model of Example 5.2 is intended to get some type of *bound* on the problem and then begin the process to narrow it while working on the solution. In the next section, a method is discussed in which a bound is created for a problem using frequency-domain analytics that allow the limits of the outcome in the time-domain to be predicted.

5.3 IMPEDANCE/LOAD BOUNDARY ANALYSIS

Now the droop in the time-domain has been examined, a method to *estimate* droop boundaries is presented, based on analytical methods in the frequency-domain. This type of boundary analysis allows the PI engineer to evaluate the efficacy of the PDN and full system in the time-domain prior to embarking on a long and arduous modeling effort. It should be made clear that this is *not* intended to replace any type of system simulation—only to give the PI engineer another tool to use when a full PDN simulation is not readily available.

The first step in the process (typically) is to find the impedance *profile* of the system based on a predetermined impedance over the frequency range of interest for the PDN [15]. The profile is strictly an amplitude plot—phase is ignored. The next step is to determine the *current* profile in the frequency-domain based on the workload estimate for the load. This is typically a spectral profile that is developed to ensure that the overall load behavior is captured over a finite time period that represents the worst-case current profile that one would see for a *semi*-infinite spectrum of current pulses.[11]

As shown in Figure 5.30, the current spectrum $I(f)$ is made up of a combination of *Fourier* components derived from the time-domain functions created by the load profile in the time-domain. The current load *function* in the time-domain consists of *a semi-infinite* summation of repetitive pulses that make up these spectral impulses in the frequency-domain, as shown in the figure. The object is to determine a profile function by which to represent the envelope of all of these impulses so that one can create a new function in the frequency-domain that is a contiguous profile of this spectrum:

$$I_L(t) = \sum_{k=1,\ N=1}^{\infty} I_k(t - NT) \tag{5.13}$$

Equation (5.13) sums the series of current pulses over an infinite period [17]. In addition, each function is shifted in time relative to the other by a multiple time

[11]This is semi-infinite because we wish to represent the periodic function as an infinite series. However, in practice, we would truncate these pulses to essentially form a basis set of finite functions.

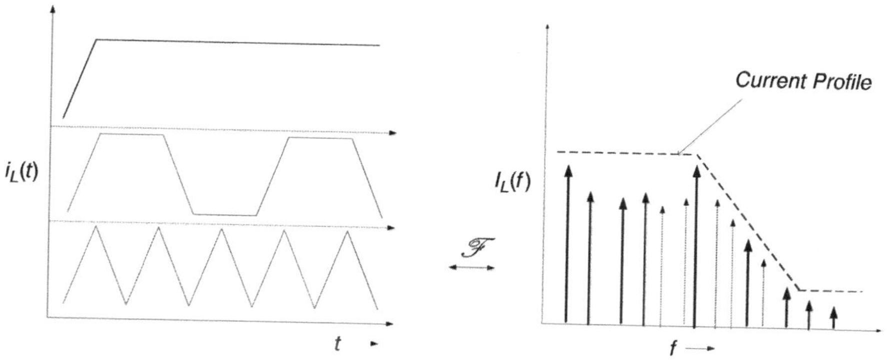

Figure 5.30 Transition to spectral profile for current function

period *NT*, which results in each function aligning end to end in time.[12] If the Fourier series of this function is taken, as stated, the result is an infinite number of impulse functions in the frequency domain. There is also an interesting aspect to the creation of the Fourier spectrum current profile with respect to some of the key elements of the time-domain pulse creation. If the period is doubled, for example (while keeping the rise-time constant), the spectral profile shifts to the left as seen in the Mathcad plot of Figure 5.31. If the rise-time of the signal is kept constant, there will be a finite limit to the envelope in the frequency domain, which essentially bounds *I(f)*. The time-dependent functions that generate the spectrums in Figure 5.31 are similar to those shown in Figure 5.30. Mathematically, they are represented as an individual infinite series. For example, the series

$$I_1(t) = c_0 a + \sum_{k=1}^{\infty} [ca(k) \cos(k\omega a t)] \tag{5.14}$$

is simply a series of triangular pulses. Equation (5.14) is in fact a Fourier series representation [16].

The profile or *envelope* is a function that tracks over the spectrum generated by the Fourier series. This is a simple variation on the *sinc* function:

$$\text{sinc}\,(x) = \frac{\sin\,(x)}{x} \tag{5.15}$$

The new function is an *esinc*, or extended sinc, function, where

$$\text{esinc}\,(x) = \begin{cases} \text{sinc}\,(x), & |x| < \dfrac{1}{2} \\ \dfrac{1}{\pi x}, & \textit{otherwise} \end{cases} \tag{5.16}$$

[12]The time duration of each pulse function is represented here as having an integer multiple-of length. In the general form, each current function would have a different length.

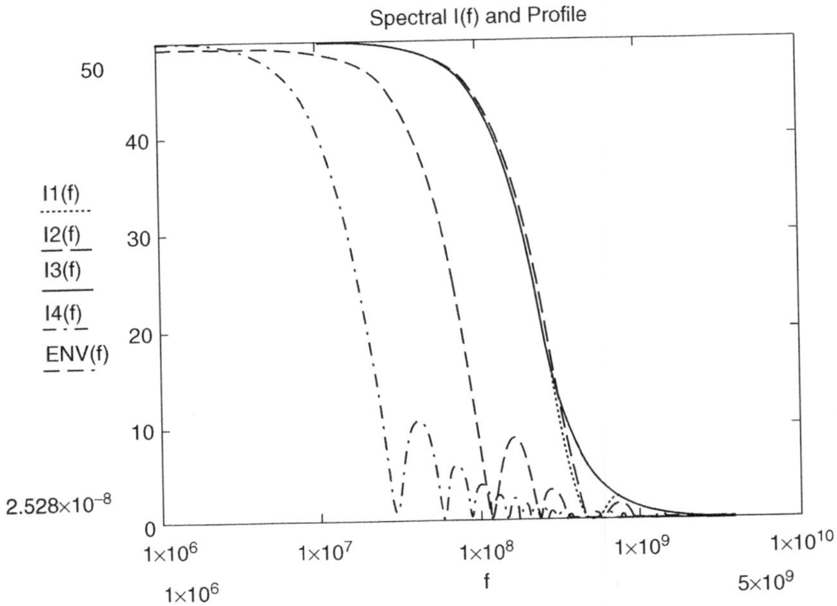

Figure 5.31 Spectral plot of various time-domain current series with profile

The current profile function is a combination of two esinc functions:

$$I_p(f) = A \text{ esinc } (\tau f) \text{ esinc } (\tau_r f) \tag{5.17}$$

which is shown for a particular pulsewidth (τ) in Figure 5.31. The reason for the two esinc functions is that one function rolls off at a different rate than the other. It has already been determined what the impedance *profile* amplitude function should look like—given the impedance in Chapter 4—which was used for the time-domain analyses in the previous section. The general shape of the impedance *profile* function is shown in Figure 5.32. This function is truncated at some point, since the bandwidth of the spectrum is limited by the edge rate of the current pulse, and the frequency spectrum for the impedance is not relevant after some frequency (~1 GHz).

The objective here is to *design* the impedance profile so that it is *convolved* with the current profile in the frequency-domain. Because there is no phase information, there can be no resonant behavior captured that limits the methodology to systems with only some level of dampening—which is the case for virtually all PDN systems. The esinc function will have phase information associated with it; however, this is different from the actual Fourier series for the current pulses and thus the correlation back into the time-domain will be inaccurate.

The esinc function gives a good starting point for the current profile. To simplify this further, one could construct asymptotic Bode diagrams of the profiles for both frequency-dependent functions, and this way get to a result more efficiently.

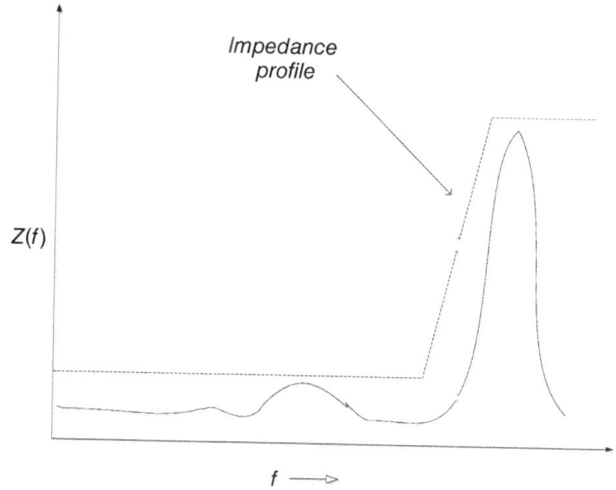

Figure 5.32 Impedance profile example

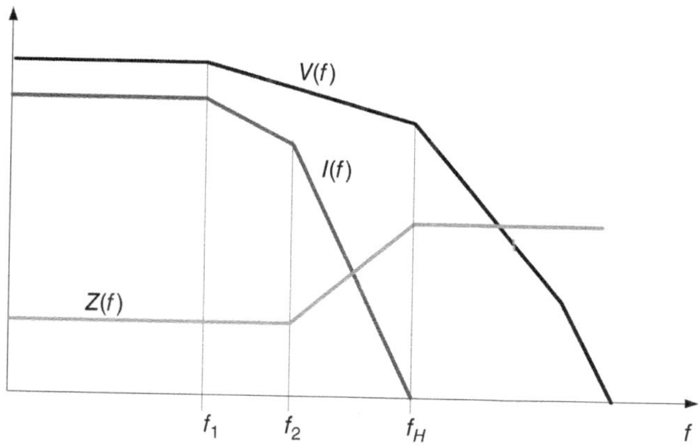

Figure 5.33 Frequency-dependent profiles for $V(f)$ profile construction

Moreover, this can be done without loss in generality. Figure 5.33 illustrates this method. From the series of profiles, the eventual function introduced is the $V(f)$ function. This is the function of interest here. In the diagram, the bounded voltage droop waveform is re-created through a construction of the impedance and current profiles in the frequency domain. The final step is to perform a *pseudo*−inverse Fourier transform of the voltage waveform, which should give the final maximum voltage droop in the time-domain—provided that the initial functions were bounded.

To ensure that the functions are bounded, it is necessary to look back at the functions for the current and impedance profiles. The estimate should give the maximum droop based on the current pulse parameters and the impedance profile of the PDN. This is illustrated in Figure 5.34.

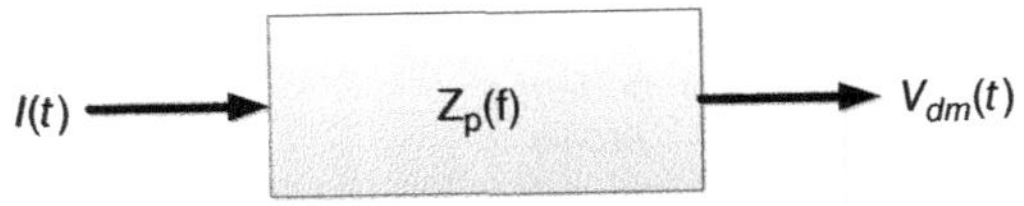

Figure 5.34 Transform method from current pulse to voltage droop

Before an example is given, it is important to know how to determine if the result is *bounded*. In other words, if one were to create a pair of profile functions in the frequency-domain, would the result be a bounded voltage-domain droop function in the time-domain? To answer this, one must first assume that both functions *are linear and time-invariant* (LTI)—that is, both functions must be LTI in order to ensure that the result will also be LTI.[13] Mathematically, the magnitude of the function $i_p(t)$ that is bounded is written as

$$|i_p(t)| < B, \quad |z_p(t)| < A \tag{5.18}$$

where A and B are bounded finite values. Thus, it is found that the function $v_{dm}(t)$ is bounded if

$$v_{dm}(t) \leq \int_0^{\infty} |z_p(\omega - \omega_N)|\ |i_p(\omega)|\ d\omega_N \leq BA \tag{5.19}$$

The convolution integral is due to the inverse Fourier transform in the frequency-domain. Equation (5.19) applies if the integrand is absolutely integrable over the time window defined. Additionally, the magnitude of the droop is generally less than v_{dm} if the maximum frequency of the impedance and current profiles is limited in the frequency domain:

$$\left|v_d(t)\right| \leq \left|\frac{1}{N}\int_0^{\omega_{\max}} z_p\left(\omega - \omega_N\right) i_p(\omega) d\omega_N\right| \leq \max\left(v_{dm}\right) \tag{5.20}$$

This is another way of stating that the maximum value for v_{dm} will always be larger than any droop event given the Fourier representations for the profile functions. Equation (5.20) is simply an inverse Fourier transform of the two functions. Note that the functions were represented under the integral with *angular frequency* for correctness, instead of *frequency*. The creation of the functions essentially gives an upper bound to the droop in the time domain. Once a basic profile is created, one may discretize the *analog* profiles and then integrate them over the frequency of interest.

$$v[n] = \left\langle \frac{1}{N}\sum_{k-1}^{N} z_p[n-k]\ i_p[n] \right\rangle_{peak} \tag{5.21}$$

where N represents the maximum discrete frequency. The goal is to examine the peak value that results in an approximate maximum value for the voltage droop. The

[13]Linear and time invariance is usually assumed for such systems for power delivery. However, it is important to make this assumption mathematically.

complexity of the method depends on the user's discretion and the desired amount of accuracy. If an actual esinc function is used for the profiles, the integration is slightly more complex. However, sometimes the PI engineer is only after a gross approximation. Thus, again, asymptotic methods could be used to estimate this droop following the same procedure. Below is a simple example of an asymptotic analysis.

Example 5.3 Compute the maximum bound of a droop event using the asymptotic profiles for the impedance and current as shown in Figure 5.35. Use f_F as the maximum integration frequency and generate the upper bound based on equation (5.20). Assume the values for the frequency intercepts as follows:

$$f_1 = \frac{1}{\pi\tau_1}, \quad f_2 = \frac{1}{\pi\tau_r}, \quad f_F = \frac{1}{\pi\tau_F} \tag{5.22}$$

where

$$\tau_1 = 10\ \tau_r = 100\ \tau_F \tag{5.23}$$

Use the values in the graph for the amplitudes, and to ease the integration process, use the following profile functions:

$$I(f) = \begin{cases} B, & 0 \le f \le f_1 \\ -B\pi\tau_1 f, & f_1 \le f \le f_2 \\ \frac{-B}{10}\left(\frac{99\pi\tau_1 f}{9} - \frac{1}{10}\right), & f_2 \le f \le f_F \end{cases} \tag{5.24}$$

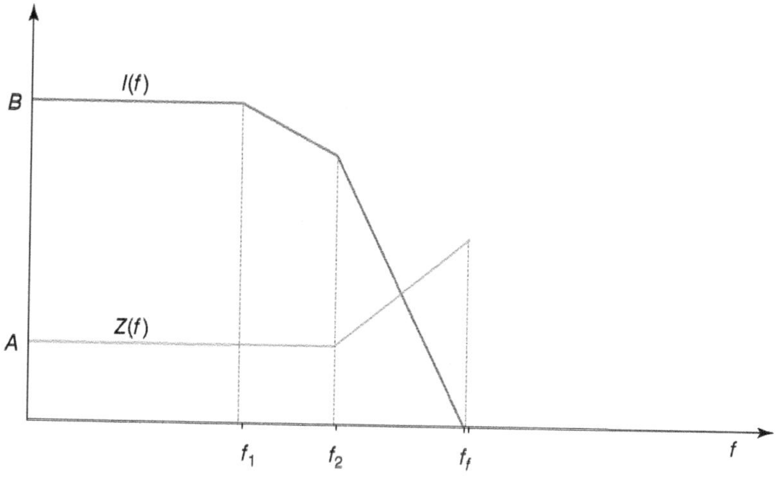

Figure 5.35 Profiles for boundary droop analysis

$$Z(f) = \begin{cases} A, & 0 \leq f \leq f_2 \\ -100A\pi\tau_1 f + 11A, & f_2 \leq f \leq f_F \\ 0, & \text{elsewhere} \end{cases} \tag{5.25}$$

Also assume A = 10 mΩ and B = 20 amps for the discrete analysis.

SOLUTION: The asymptotic profiles are mathematically represented as two functions multiplied together in the frequency domain analytically. The goal is to *sum* over the three regions after discretizing the profiles. First, use equation (5.21). The next step is discretization. By sampling every decade, a crude approximation across the spectrum can be achieved. The amplitudes here are based on those set in the time-domain—therefore, the mathematics in the discrete frequency domain is quite simple—all one needs to do is perform a convolution to get the result and then divide by the number of samples. Figure 5.36 gives the discretization for the estimates.

A spreadsheet program computes the values for $v[N]$. The convolution is a simple shift of the elements multiplied by each N element across the discrete spectrum. Because the multiplication is by each an independent value, the computation is just a sum of the convolved elements multiplied by the nonconvolved elements. The results are shown in Figure 5.37.

If one shifts the impedance profile downward in frequency, the maximum value is expected to increase. Here, the maximum value occurs below roll-off points. For some profiles, this may occur at a higher frequency. Figure 5.38 shows where the impedance profile was shifted downward and the maximum estimated droop increased.

The data indicate that the droop events should not exceed 250 and 350 mvpp, respectively, for the given constraints. However, these are still crude estimates and

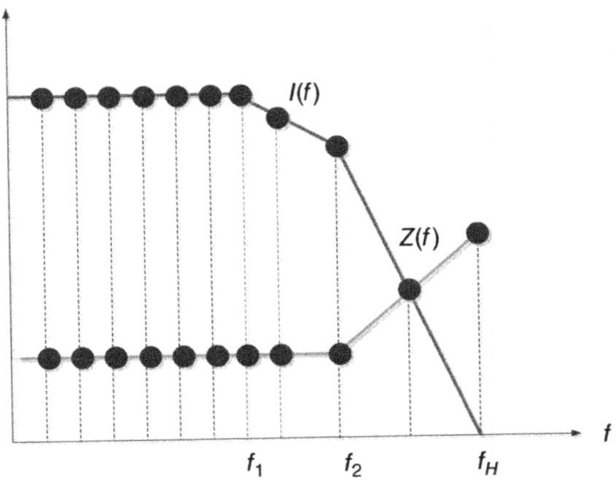

Figure 5.36 Discretized version of profiles

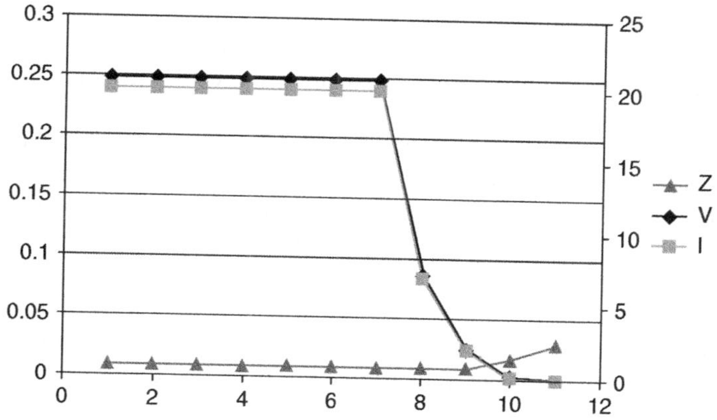

Figure 5.37 Discrete convolution results showing I, Z, and V

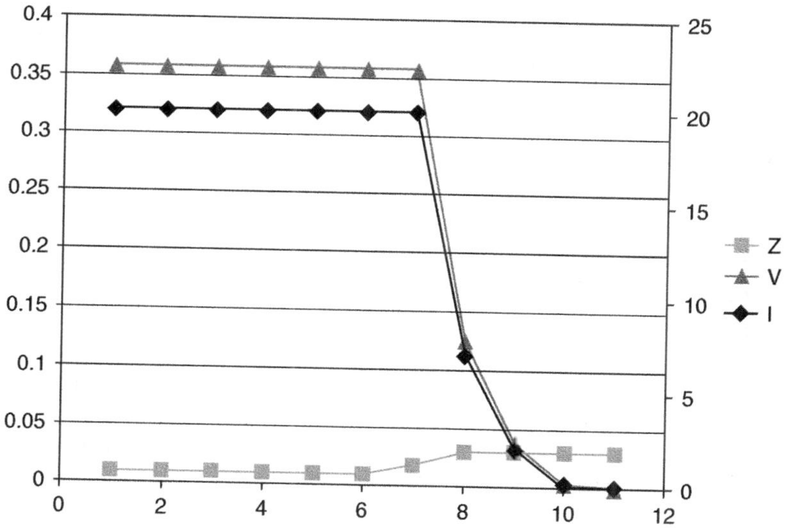

Figure 5.38 Shift of impedance profile to left and results for V

should not be relied upon to give the final answer for modeling effort. The spreadsheet is a quick and simple way to determine if the droop events will be larger than might be desired. Whether using a crude PDN in the time-domain or using this method, the objective is always to gain insight into the problem prior to modeling it.

5.4 SUMMARY

In this chapter, methods for modeling were examined for both sources and loads for time-domain analyses. The focus was on the buck converter as a source model, and both open-loop and closed-loop source models were discussed. The reason for including both models is that sometimes PI engineers do not need the sophistication of an advanced converter model to analyze the voltage droop of the system. There were also presented a number of methods toward modeling the load, the most important being the G model in SPICE. The source of the load at the silicon device was also considered and focused on a microprocessor to show how high di/dt systems would behave. Among the key parameters, the slew-rate of the step-function was reviewed. Time-domain droop events were then used to illustrate how these droop events are characterized in time—in relation to the frequency-domain analysis of the previous chapter. Each droop event is unique based on the activity at the load and how the silicon device responds to these loads. An example was given on how to integrate the PDN with the source and load, and then some basic results were shown with two different source models, which illustrated that the difference is mainly in the low-frequency response, as expected. Finally boundary value analysis in the frequency-domain was discussed, and how to determine a priori what the droop will be based on knowledge of the load profile and the impedance distribution profile. This simple method enables a quick estimate to be made of the droop event using a crude discrete Fourier analysis.

REFERENCES

1. Paul, C. R., and Love, C. A brief SPICE tutorial. *IEEE EMC Society Newsletter*, no. 225, Spring, 2010.
2. *PSPICE Users Guide*, Cadence Design Systems, http://www.cadence.com, Inc., 2009.
3. *PSPICE Reference Guide*, Cadence Design Systems, http://www.cadence.com, Inc., 2009.
4. Kasat, S. Analysis, design, and modeling of DC-DC converter using simulink. Masters Thesis. Oklahoma State University, 2004.
5. Griffin, R. E. Unified power converter models for continuous and discontinuous conduction mode, analysis. *PESC*, pp. 853–860, 1989.
6. McDonald, K. F. *Dependent source modeling for SPICE. IEEE Pulsed Power Conf.*, pp. 365–368, 1991.
7. Luo, Y., Dougal, R., and Santi, E, Multi-resolution modeling of power converter using waveform reconstruction. *Simulation Symp. Proc.*, pp. 165–174, 2000.
8. Smedley, K. M., and Cuk, S. One-cycle control of switching converters. *Power Electronic Specialists Conf.*, pp. 888–896, 1991.
9. Ninkovic, P., and Jankovic, M. Tuned-average current-mode control of constant frequency DC/DC converters. Industrial Electronics Symp., pp. 235–240, vol. 2, 1997.
10. Veneable, H. D. The K-factor: A new mathematical tool for stability analysis and synthesis. *Powercon* 10, pp. H1–12, 1983.
11. DiBene II, J. T. Integrated power delivery for high performance server based microprocessors. PwrSoC, 2008.
12. Hochanson D. H., and DiBene II, J. T. Power delivery for high performance microprocessor packages—Part I. Electromagnetic Compatibility: *IEEE Int. Symp*, 2007.

13. Liu X., and Liu Y-F. The extraction and measurement of on-die impedance for power delivery analysis. *IEEE EPEPS*, 2009.
14. Bowman K., et al. Dynamic variation monitor for measuring the impact of voltage droops on microprocessor clock frequency. *IEEE CICC*, 2010.
15. Meijer, M., Pessolano, F. and Pineda de Gyvez, J. Technology exploration for adaptive power and frequency scaling in 90nm CMOS. *IEEE ISLPED*, 2004.
16. Paul C. R. *Introduction to Electromagnetic Compatibility*. Wiley, 1992.
17. Oppenheim A. V., and Willsky A. S. *Signals and Systems*. Prentice-Hall, 1983.

PROBLEMS

5.1 Determine the asymptotic bode plot poles and zeros using the data from Example 5.1.

5.2 Compute the compensation values using Veneable's K-factor method given in section 5.1.1.

5.3 Develop the schematic for Example 5.2 and run the closed-loop switching model with the same impedance values as given in the example.

5.4 Increase the capacitance by 5× throughout the PDN and the voltage regulator in Problem 5.3. What is the maximum third-order droop?

5.5 Set up the third integral for Example 5.3 and solve it. Is this the dominant maximum droop?

5.6 Create a spreadsheet similar to the one described in Example 5.3 using the same data. What is the droop event if the peak impedance is twice that in the example?

6

SYSTEM CONSIDERATIONS FOR POWER INTEGRITY

Besides the power source, load, and distribution elements, there are a number of platform-level elements that need to be considered (electrically) when analyzing power at the system level. As observed in the preceding chapters, not only does one need to consider the PDN but also how the PDN responds to the dynamic behavior of the load and source and the efficacy of the power that is delivered to the load. In the time-domain analysis, of Chapter 5, it was clear that load perturbations indicate how good (or bad) the power distribution path and full system power delivery is. This was done by analyzing the different voltage droops due to different loads. However, only the simple case was examined where the voltage *setpoint* was kept static. What if this setpoint were allowed to move in response to load changes? This is one of the areas that is explored in this chapter to broaden the scope of the power integrity engineer's learning. Also the effects of noise in the PDN will be introduced, since this adds to what is termed the *guardband* of the power system. The different noise sources will be broken down and examined in terms of how they affect the quality of the power distribution network at the load. Also considered will be the different techniques used to squelch some of the noise sources. We examine electromagnetic interference as it pertains to power integrity, different EMI sources, and emissions due to EMI. Finally, some basic measurement techniques for looking at PDN-related power integrity will be discussed.

Fundamentals of Power Integrity for Computer Platforms and Systems, First Edition. J. Ted DiBene II.

6.1 POWER LOADLINE FUNDAMENTALS

The key to good power integrity is to ensure that the power source is stable during all perturbations at the load and that the *data integrity* is maintained throughout any and all droop events. Many advanced silicon devices require that the voltage *set-point* change in order to alter the frequency and performance of the system during normal operation. This was discussed briefly in chapter 5. When the voltage set-point is increased, the computing device will operate at a higher frequency allowing for higher performance and *through-put* from the processor. However, what goes along with this higher bandwidth and performance is larger droop events during the processing of the workloads. This is an expected result, since the device typically requires more current in a shorter amount of time to process the data at a higher bandwidth. In some cases, what can happen is the droop event may *increase* beyond the power systems and PDFN's ability to deliver power adequately or, it requires the PI engineer and the power converter designer to overdesign the system resulting in a more costly solution. This is certainly not required of all power rails supporting the silicon device, but for high performance computing and processing, it is the case.

One way to combat this problem is to increase the voltage setpoint higher than nominally required, thus yielding additional voltage guardband against spurious droop events. Nonetheless, there is a risk associated with such an approach where the result could be damage to the device if the setpoint is placed too high. Moreover, the higher setpoint results in higher power loss for these devices.[1]

Figure 6.1 shows what happens if the voltage setpoint is set too high, or conversely, set too low. At the left side, the voltage setpoint is set at too low a point, which results in the first-order droop, or *undershoot*, dropping will below the

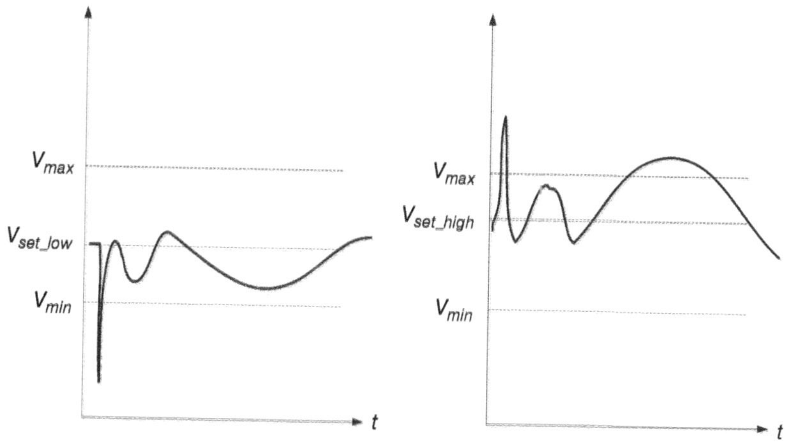

Figure 6.1 V_{max}/V_{min} and voltage setpoint issues

[1]Higher voltages in nanometer technologies can increase leakage current. The transistor leakage is a function of voltage (among other parameters) and increases exponentially with the voltage.

minimum allowable voltage for the device. This can occur also when the droop event is larger than the system response. At the right side, the voltage setpoint has been set *too high*, resulting in an *overshoot* event that exceeds the maximum voltage that the device can handle. The silicon device can then be damaged if the voltage exceeds the acceptable reliability parameters for the device and process. As silicon device voltages decrease, the maximum and minimum thresholds for them shrink, potentially causing operational errors or failures. A way around this problem is the *loadline* or, as has been discussed previously in Chapter 2 (though briefly), using *adaptive voltage positioning* (AVP) in the converter design.

6.1.1 Loadline

The *loadline* in a power delivery system is a function of the DC resistance for the PDN that is designed for that particular silicon device (see Figure 6.2). The DC resistance of the path takes a voltage drop, as seen at the load, and thus there is always a difference between the voltage sourced by the VR and of that seen by the silicon.

For a given current, the voltage differential will increase relative to the DC loadline. However, to guarantee performance, the silicon requires that the voltage be within a certain region, and under specific operating conditions. Thus, in the feedback path, the voltage regulator needs to adjust the voltage to maintain the correct value over the working conditions of the device. Figure 6.3 shows the *nominal*

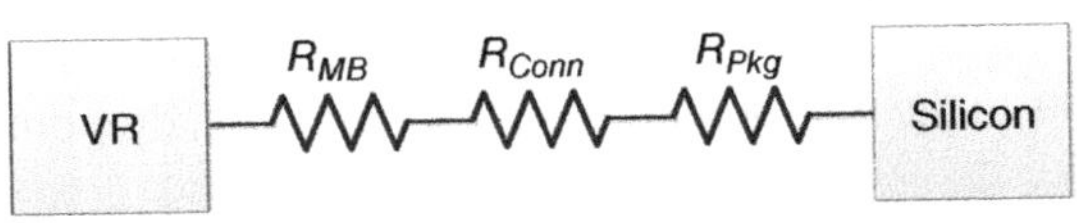

Figure 6.2 DC distribution path of PDN

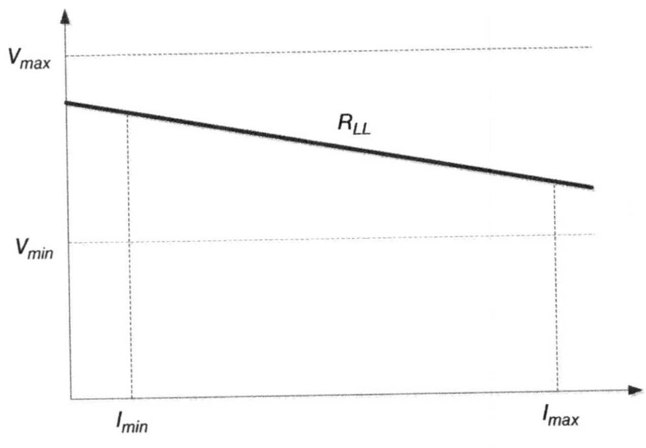

Figure 6.3 Basic loadline

loadline for a basic power delivery scheme. The term *nominal* means that there is no *error* associated with the loadline in the figure. In general, there will be an error or tolerance around every practical loadline design. These errors are addressed later in the chapter.

The operation of a loadline is simple, As the average current in the load moves from a lower value to a higher one, the voltage drops. However, the voltage at the load must still be within the regulation limits for the silicon device to operate correctly. This operational region is predetermined by programming the VR, to guarantee, that all voltage limits will be met across the loadline. For illustration purposes, it is assumed in Figure 6.2 that only one voltage *setpoint* has been set.

When current increases at the load, the VR will force the voltage to adjust to the proper value. This means that the voltage regulator must be able to *sense* the current as well at the output voltage to ensure that the proper output value is maintained. If a sudden step change occurs in the current—for example, the unload in this case—the voltage readjusts to a point at the load where the overshoot will not damage the device (see Figure 6.4).[2] By definition, a current droop event cannot occur at $I_{\max}$ because this is the maximum current the load can create under the given operating conditions.

The programming of the loadline is done through the communication port that connects the silicon device to the VR. For microprocessors, this port is called a *voltage ID* (VID) port. It is parallel or serial in design, but its function is the same—to send messages to the VR to adjust its voltage to the correct setting. The VID setting is determined at a minimum or *zero* load. In practice, the loadline has multiple VR settings, and thus, multiple *loadlines*. This is because for every operational frequency of the device, a different voltage is set to ensure proper operation and performance.

The loadline likewise has a tolerance associated with it. This tolerance is due to the *error* that may occur from the sensing and measurement of the DC resistance

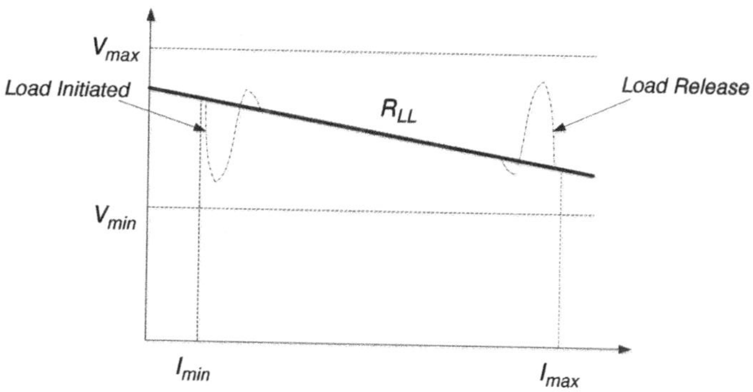

Figure 6.4 Loadline with droop and overshoot events

[2]Dynamic behavior is not shown in loadline curves because the loadline is, by definition, a very low frequency or DC event. The dynamic voltages are shown here for illustrative purposes only.

of the path, which is difficult to measure accurately once a real system is built. The question that typically comes up in discussions of loadline behavior is how fast is the response of this loadline in a real system? In practice, the measured current data are fed back in real time along with the sensed voltage data. However, the current sense data are typically slower than the voltage sense data, and thus, the response is slower than the total loop response of the VR. This is, however, considered adequate for most power delivery systems, since the time delay in compensating along the loadline is not an issue if the loadline and initial voltage setpoints are set correctly. The loadline is considered a *static* event, and thus, does not rely on the PDN's-high-frequency filtering to control its behavior.

6.1.2 Tolerance Band and Voltage Guardband

For a given power delivery scheme, there are practical offsets and errors within the measurements. Many of these are associated with the VR controller—some have to do with the filtering of the converter and the PDN. As stated in the previous section, the loadline also has errors associated with it, which can be seen in Figure 6.5.

Note that the slope of the loadline is a function of a number of factors. Depending on the type of current sensing the VR employs, the effects can be large or small. For sensing across the storage inductor, a number of these effects are listed below:

- L&R temperature of inductor—inductance and DC resistance.
- L&R tolerances—static manufacturing tolerances of inductor.
- R&C tolerances—error in RC network across inductor.
- Op-amp resistors—errors in the resistor network feedback (typically resistor mismatch) to current sensing op-amp.
- Op-amp—random offset error and gain error.

A simple schematic for the current sensor is shown in Figure 6.6. The slope of the line changes as a function of the current load for the error. This is because

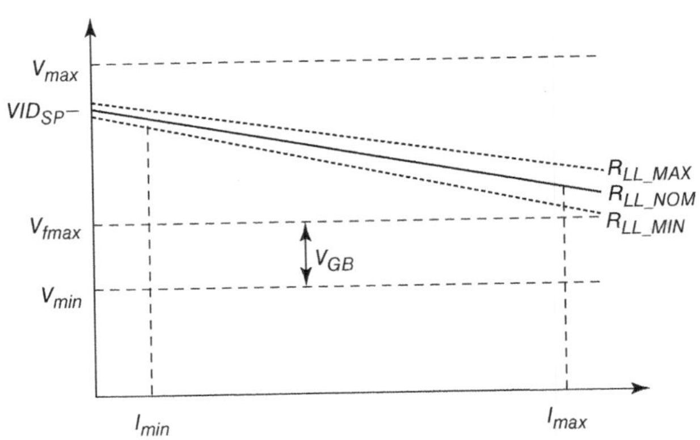

Figure 6.5 Loadline tolerance, VID setpoint, and voltage guardband

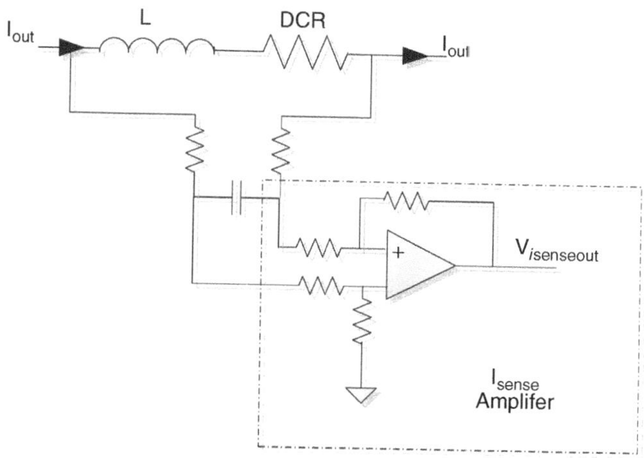

Figure 6.6 Current sense amplifier example for inductor sensing

the errors that accumulate in the sensing path directly influence the value of the loadline that is expected [3]. Because the error is constant, but multiplies with the current, the error increases with the load. There are additional errors associated with *voltage* sensors as well. Typically, however, these are smaller errors than those of the current-sensing devices.

To measure the loadline with the sensors shown in Figure 6.6, it is necessary to measure the output voltage at the source and the load along with the output current, or mathematically,

$$R_{LL} = \frac{V_S - V_L}{I_{Out}} \tag{6.1}$$

where R_{LL} is the loadline resistance. The errors associated with the loadline accumulate, and are typically statistical in nature. For example, the errors in the op-amp offset are dependent on a number of factors, including the possible calibration methods used to reduce them. The resistors in the feedback path may also have errors that are due to manufacturing tolerances and are harder to compensate for—particularly if the components are discrete and not part of the silicon device. The error accumulations affect the tolerance of the loadline in equation (6.1). Thus, the *change* in the loadline resistance is due to the changes in each variable that is measured:

$$R_{LL} \pm \Delta R = \frac{(V_S \pm \Delta V_S) - (V_L = \Delta V_L)}{(I_{Out} \pm \Delta I)} \tag{6.2}$$

As the loadline values drift due to these errors, the overall *tolerance band* (TOB) also changes [1]. The TOB for the system is a measure of the total error in the power delivery system—including the voltage setpoint accuracy across the total loadline—where the loadline *slope* errors are included. The *typical* TOB error sources are shown schematically in Figure 6.7.

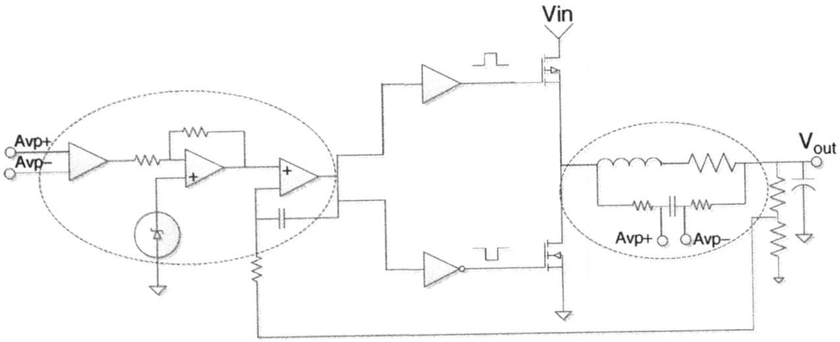

Figure 6.7 TOB typical error sources (circled) in a simple buck converter example

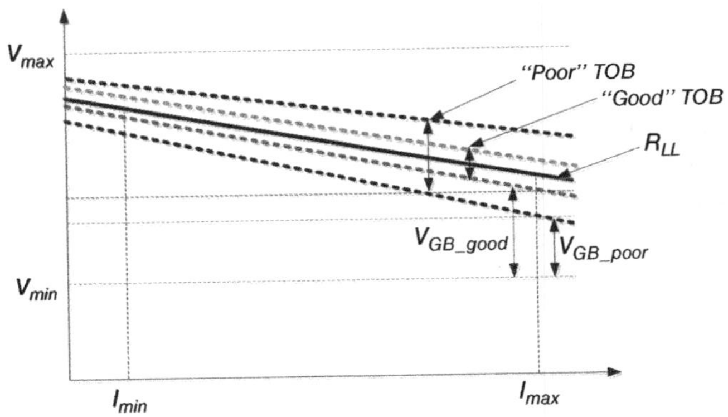

Figure 6.8 TOB defined with voltage guardband

The errors for the loadline tolerances dominate and accumulate in the sensors (amplifiers) along with the reference data and the filter tolerances [2]. Figure 6.8 shows the TOB along with the defined voltage guardband. For illustrative purposes, both a *good* and a *poor* TOB and voltage guardband are shown. The good TOB is obviously one where the errors around the nominal loadline are minimal. Thus, this leaves more room for the voltage guardband in the system. The poor TOB takes up a larger portion of the available margin between V_{min} and V_{max}, and thus, limits both reliability and performance in the system.

Understanding all aspects of every error tolerance is beyond the scope of this book, and the reader is encouraged to review the references at the end of the chapter. It is the intention here to provide insight into the key tolerance elements and how to determine the contribution due to the total, so that one can better understand where the margins are for the system. Once these margins are known, the voltage guardband is determined. For the PI engineer, this is key, since the objective is to ensure the integrity of the power that is delivered under all conditions.

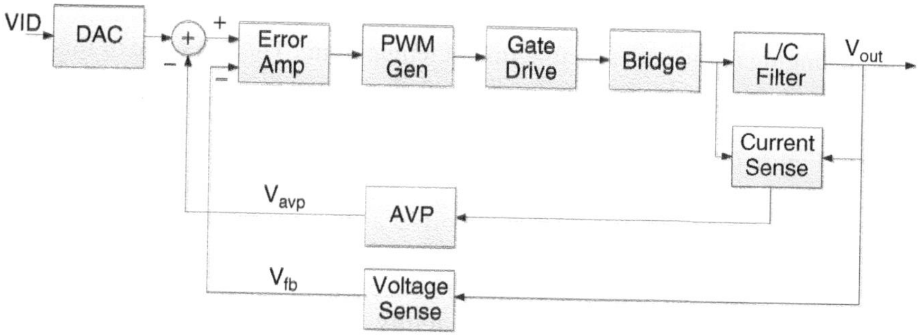

Figure 6.9 Block level diagram for current and voltage sensing and measurement

Figure 6.9 shows the full path block diagram for the feedback. Understanding this path allows the comprehension of the errors and their contributions. Much of this material was previously discussed in Chapter 2—without all the details, of course—here, the feedback is examined with respect to TOB. Starting at the left-hand side of the diagram, the VID settings come into the DAC (digital-to-analog converter) in the form of digital bits, which represent the analog voltage setting for the VR. The DAC converts these digital bits to analog where the analog signal is fed back into the error amplifier at a summing junction. Along with the DAC analog voltage, both the *converted current* sense voltage and the *voltage* sense are fed back into this junction. The purpose is to adjust the output voltage to the proper setpoint given the reference data that has been delivered by the DAC and the sensors. As seen in previous chapters, the error amp will compensate for any difference and adjust the value accordingly. The error amp has its own voltage reference—as discussed in Chapter 5, which is typically a *bandgap* reference—that is used to compare it against the other inputs. The other components have been discussed in detail in previous chapters, mostly in Chapter 2.

The goal, as stated, is to determine what the total TOB is for the loop. The TOB comprises the manufacturing tolerances (just discussed), the output voltage ripple, and the voltage offset due to temperature. Or put another way,

$$TOB = TOB_{mfg} + V_{ripple} + V_{TT} \tag{6.3}$$

The TOB typically includes the ripple and thermal voltage offset, since these elements often result in the biggest component limiting the voltage guardband. The voltage ripple is simply the ripple as measured by the tester and determined by the specification limits set within the system power delivery scheme. This is the ripple as seen *at the load.* The voltage ripple is a function of the converter operation as well as the PDN filtering. Because the ripple is a low-frequency event, relative to the silicon device switching, it essentially directly contributes to the total the TOB.

The thermal drift is the voltage offset that is due to thermal variations in the system over the normal temperature operating ranges. There is more than one component in the TOB path that is affected by the thermal differences in the system.

An example of a simple component of this is the *drift* in the DC resistance of the copper changing in the path as a function of temperature. Though some of this drift should be compensated for in the feedback path, there will always be some offset that accumulates as an error. The TOB accumulation from all three components is shown in Figure 6.10, where the top graph at the right shows the contribution due to all the elements.

The voltage ripple and voltage thermal offset accumulations are typically specified quantities and accumulate directly into the overall TOB. The TOB_{mfg}, however, accumulates over a statistical variation. There are a number of ways designers and manufacturers determine the range of variability, and each organization may treat it differently. A common method is to use the *root-sum-of-squares* (RSS) method to determine the TOB variations. The total distribution is the sum of the errors accumulated in the path

$$\sigma_{total} = \sqrt{(\varepsilon_1)^2 + (\varepsilon_2)^2 + \ldots + (\varepsilon_{n-1})^2 + (\varepsilon_n)^2} \tag{6.4}$$

where ε refers to a given error associated with the guardband computation. The variable σ we know through statistics as the *standard deviation* from the *mean* value. One needs only to convert the analysis from a generic standard deviation to *millivolts* for the total offset from the nominal, as shown in Figure 6.5. The new equation is

$$TOB_{mfg} = \sqrt{(VID \cdot N_{VID})^2 + (V_{AVP})^2 \cdot \left(N_{CE}^2 + \frac{N_{ESR}^2}{n_{ph}}\right) + (V_{AVPdyn})^2 \cdot \left(\frac{N_L^2 + N_C^2}{n_{ph}}\right)} \tag{6.5}$$

where the coefficients are as defined in Table 6.1.

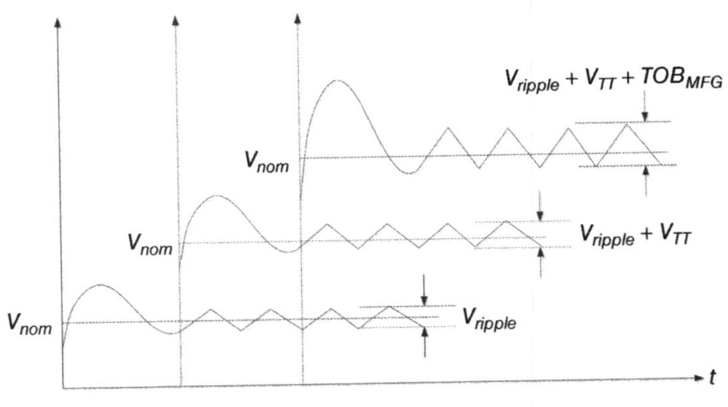

Figure 6.10 Error accumulations in TOB

TABLE 6.1 Error values for TOB_{mfg} computation

Coefficient/Variable	Definition
VID	VID voltage value—set at nominal voltage for error analysis
N_{VID}	Percent error on VID setting—determined by DAC error
V_{AVP}	$I_{max}*R_{LL}$—the total adjustment voltage ΔV across the nominal loadline
N_{CE}	Current sense error in controller as a percent error
N_{ESR}	Percent error on ESR of inductor
n_{ph}	Number of phases for VR
V_{AVPdyn}	$I_{dyn}*R_{LL}$ = total voltage drop for dynamic step change
N_L	Percent error in inductor filter (VR)
N_C	Percent error in capacitor filter (VR)

Many of the coefficients are given in terms of the total deviation (2σ, 3σ, etc.) to get the overall tolerance band. It should be noted that equation (6.5) is an overall *estimate* for the manufacturing tolerance band. The PI engineer or the VR designer may choose to compute a more detailed study that includes some additional guard-bands to obtain a more exact estimate. The next example illustrates the usage of the equation with a simple computation and analysis.

Example 6.1 Compute the total tolerance band for a power delivery system given the values in Table 6.2. If the voltage ripple is 10 mVpp, and the voltage thermal drift is 5 mV, what is the total tolerance band? Assume the total tolerance band is based on three standard deviations. What would the tolerance band be if the designer wanted a 5σ distribution?

TABLE 6.2 Guardband values for computation

Coefficient/Variable	Metric
VID	1.25 V
N_{VID}	0.5%
V_{AVP}	90 mv
N_{CE}	2%
N_{ESR}	5%
n_{ph}	4
V_{AVPdyn}	$I_{dyn}*R_{LL}$ = 75 mv
N_L	15%
N_C	8%

SOLUTION: First, use equation (6.5) to compute the 3σ deviations. It is a matter of plugging in the values

$$TOB_{mfg(3\sigma)} = \sqrt{(0.00625)^2 + (0.9)^2 \times (1.02e-5) + (0.75)^2 \times (7.2e-3)}$$
$$\cong 9 \text{ mVpp} \tag{6.6}$$

If the total guardband is added to the equation, the result is

$$TOB = 0.009 + 0.010 + 0.005 \cong 24 \text{ mVpp} \tag{6.7}$$

This is a reasonable guardband for a high-performance power delivery system. There are times, however, when it is desirable to increase the number of standard deviations in the guardband to ensure that the yields are also high for the platform, so that there is less opportunity for failure in the field due to the guardband errors. If one now increases the number of standard deviations to 5, the result becomes

$$TOB_{mfg(5\sigma)} = \frac{5}{3}\sqrt{(00625)^2 + (0.9)^2 \times (1.02e-5) + (0.75)^2 \times (7.2e-3)}$$
$$\cong 15.6 \text{ mVpp} \tag{6.8}$$

which yields a total TOB of around 31 mVpp. Though this does not sound like a large increase, from a cost perspective, for lesser returns from potential field failures, it could be significant. The penalty, of course, would be there would be less voltage guardband (as will be seen later) and possibly lower performance for the device. It is often necessary to make some difficult choices between selling a highly reliable system and one that has the top performance under all conditions.

Now that TOB has been discussed, it is appropriate to return to the discussion of voltage guardband. This is essentially, what is left over when TOB is subtracted out from the loadline at I_{max}. This voltage differential is often one of the key elements to performance in a system as was discussed above in the example. The voltage guardband is a predetermined value once the margins for the power delivery system have been measured and assessed. The voltage guardband is intended to maximize this performance under all conditions over the life of the product [4,5]. The voltage guardband is tied to the minimum voltage within which the device can operate correctly. Since, for each operating mode, there is a corresponding loadline and V_{min} for the device, for a given VID setting, there is a specific range of operation that is different from any other range. This is seen in Figure 6.11. The operational range changes for the loads, owing to the performance-to-load current being highly dependent on the frequency of operation, and thus, the bandwidth of the device.

Notice in Figure 6.11 that the VID settings for three different values indicate three different loadlines and three different minimum voltages. At a lower VID setpoint the operating frequency is typically set lower, and thus at a lower performance level for the device. However, while the range of the current load may be different at a lower V_{min} setting, the silicon device may continue to operate correctly, even if the voltage setpoint is lower than that set for another VID setting. This is because as the switching speeds for the device slow down, the timing for the transistors

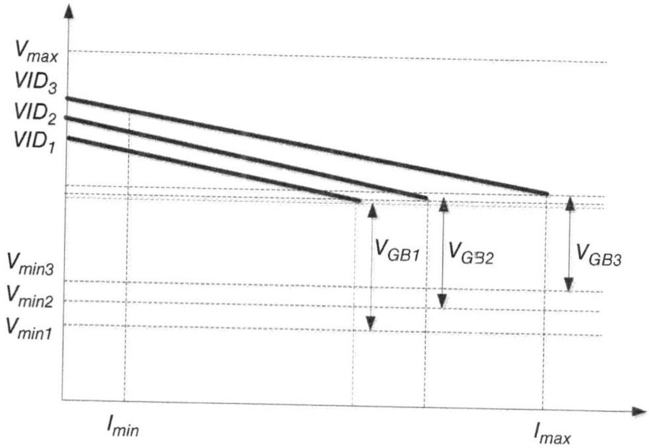

Figure 6.11 Guardbands for different VID settings

and gates of the device becomes less constrained. Conversely, as the VID settings increase, the load range increases, but so does V_{min}, resulting in a smaller voltage guardband [6].

The minimum voltages and voltage guardbands are *tested* or measured to a standard set of values. Most high-performance silicon devices comply with different process metrics established at manufacturing and are categorized according to their performances under those measurement conditions. When the device is placed in a tester, the voltages are set and measured at specific points on the package and/or die. The device is then run through a series of tests to determine its performance. To ensure the device operates within its specified range, the aforementioned guardband (a variance of) is then assigned to it. Some devices are moved into different *bins* and categorized within a range of operating speeds for a standard applied voltage. Within these bins, the devices are intended to perform within a standard set of operating metrics, including voltage. To ensure that all devices in a bin operate correctly over their lifetimes, the part of the voltage guardband is assigned to them. This way when the voltage is applied in the system, all devices in that bin will perform essentially the same. Thus, a percentage of the total guardband is applied to the overall voltage guardband computation.

Another part of the guardband computation is that of the guardband of the *tester*. The tester has limitations and inaccuracies as well, and this is added as part of the voltage guardband offset from normal operation. Though it is not shown explicitly in Figure 6.11, this guardband is implied. Typically, a manufacturing engineer will know this information for a given setup and device. Table 6.3 lists the types of elements, that contribute to a voltage guardband computation—though not in any particular order or magnitude.

The tester guardband is a function of the measurement error from the production tester used to determine the actual voltage at the silicon load over the VID range settings. This is a statistical value, and computing the absolute error is dependent

TABLE 6.3 Voltage guardband common contributors

Component	Definition
Tester GB	Production tester error (statistical)
Age, temp, and VID compensation	Effects due to aging (end of life), temperature variations, and VID offset errors (binary)
Local silicon offset	Nondeterministic offsets at local silicon due to measurement error localities
Process GB	Process variances for device

on many factors, each of which is beyond the scope of this book. The important thing for the PI engineer to be aware of is that this guardband exists and affects the overall power delivery of the system.

The next guardband that is considered is often lumped together. This is the aging, temperature, and VID compensation offset guardband. Typically, these errors are small and their contributions are less than 10–15% of the overall voltage guardband computation. Guardband errors due to aging often come in with respect to both the load device aging and the VR controller aging. As the silicon ages, there are some non-deterministic effects which are not predictable, this involves the references in both devices as well as the offsets for some amplifiers. Errors due to aging are provided by manufacturers and then added to the overall guardband. Temperature changes also add to the guardband, but even these errors change over time. As a matter of fact, there are two components here—*deterministic* and *nondeterministic* temperature effects. The first is known about the device, such as temperature drift in the bandgap reference or offset in the compensation network and op-amp in the VR. These guardband errors are fairly predictable as *start-of-life* for a product and the VR designer has an understanding of the contribution of these elements. Once the components are loaded into the board, and as the unit is operated over time, these errors drift due to temperature fluctuations and aging. However, the aging and temperature changes can often be orthogonal to each other—meaning, as to the direction the errors drift—thus, they are considered separately.

The final guardband contribution is strictly a function of the device itself and comes into play for more complex parts with multiple power sources and loads—such as the multicore microprocessor. Most platform designers—and package designers—do not have the luxury of having sense points at every contact on the silicon device. In fact, there is only one point per rail where they are able to sense the voltage, but this is only considered an *average* measurement point. Thus, while this makes it simpler for the VR and system designer, it does cause some ambiguity as to what the actual voltage is at any one position on the die. If there are multiple processing units or cores on a single die, each one will have its local ground reference and voltage plane, though all are tied to a single voltage rail. This way a voltage drop from this measurement point will be sensed relative to where this processing unit resides on the die. For most silicon devices, this may not be an issue. However, for high-performance devices this guardband must be added to the overall guardband voltage error to ensure that all the processing units operate correctly over time as well.

The power integrity engineer is not required to scrutinize every error or guardband in a system. However, having a basic understanding of these components helps the PI engineer maintain the power integrity of the system.

6.2 NOISE GENERATION CONSIDERATIONS IN POWER INTEGRITY

So far, we have covered how the loadline affects the power delivery system, and in particular, voltage guardbands and advantages (and some disadvantages) in using a loadline. Two types of noise in the power delivery system are now addressed: self-generated power bus noise and coupled power bus noise. The first type is inherent to the operation of the PDN, the converter, or the load (or any combination of the three), and this noise reduces voltage guardband in the system. Its frequency content must be kept below 1 Ghz (to protect the voltage guardband). The other noise source is due to *coupling* into the power buss from an external source. This noise source is usually caused by another power network, or signal(s), in close proximity to the power network of interest. Such coupling is common in complex PDN systems and is similar to *cross talk* in IO transmission line problems. Fortunately, when the noise is mitigated (near the source) in one area, it is often mitigated in the rest of the area. The interest in examining noise sources is how they reduce the guardband in the system. Later in the chapter, we will consider how noise from the power delivery system affects the rest of the board and chassis. First, we will look at the causes of noise and some basic methods used to model noise.

6.2.1 Self-generated Power Bus Noise

There are two sources of power bus noise in a power delivery system: noise from the *switching converter* and noise from the *load*-switching operation. The focus here will be on converter-generated noise, followed by a discussion of general silicon load noise later in the chapter.

The converter noise is primarily due to switches of the converter, with the spectral content of this noise due to the number of components that make up the switch power path. Recall from Chapter 2 the discussion of the behavior of the CMOS switches: there it was evident that even when *off*, the switches could generate noise.[3] This noise was due to the energy storage in the gate-charge of the FET switches essentially *releasing* their energy into the load after switching. For *hard-switched* topologies, this is a typical problem (see Chapter 2). Figure 6.12 shows potential noise emitted from the switching behavior across a FET switched in a mode where the energy in the FET causes ringing back into the input voltage line.

[3]"Off" is a relative term here. Shortly after the device has been turned off, capacitances in the device are still charging and discharging, emitting noise across the device and surrounding components.

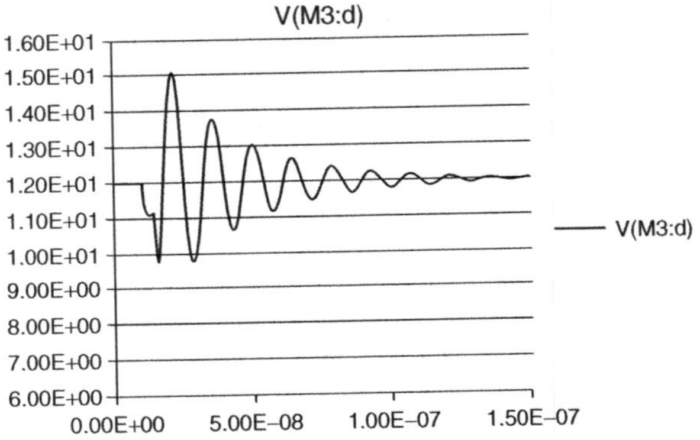

Figure 6.12 Power FET ringing example

Because of where the noise is located, most engineers assume that this will have little to no effect on the guardband. However, this is not the case. Because the duty cycle of the regulator is a ratio of the output to input voltage, the PWM signal will be affected. As the voltage droops or overshoots on the input, the duty cycle will *dither*, resulting in a change to the output voltage. This change will have an effect on the guardband. This ringing (potentially) will also show up on the *ground* return for the full return path of the bridge FET pairs (see Figure 6.13).

The current in Figure 6.13 will circulate in the duty cycle path shown and then commutate in the other loop. In the duty cycle, during the switching on the edges

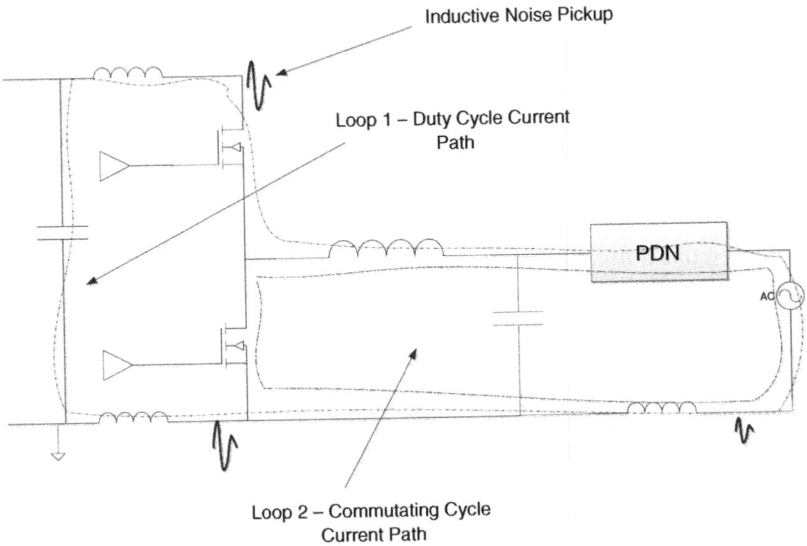

Figure 6.13 Current path with noise pickup locations

(rise and fall times), the current will flow through the parasitic inductances as indicated in the figure. Through the PDN side of the structure, there exists a ground return inductive path that is finite, which can be *driven* by this voltage source [15]. If that happens, this noise will also be seen on the *ground* side of voltage measurement across the output. For the PI engineer, this could be a significant problem—particularly if the noise is not filtered by the PDN or if the rest of the design of the converter layout is poor. What typically causes such problems is that the energy in the FET capacitance resonates with other parasitics in the system, which causes the ringing to take place [13]. Moreover, this ringing could also result in poor efficiency of the converter if it is left unchecked. If the ringing is very significant, then the line voltage is affected and the device will not turn off or on completely, resulting in more loss due to the duty cycle's dithering [14].

To mitigate the noise, it is usually left to the designer to properly layout the converter relative to the layout of the PDN. However, this is where the PI engineer and the VR designer need to work towards the same goal; delivering high quality power delivery to the device being powered. That is to say, their objective would be to minimize the loop inductance in the layout for the devices so that the FET switches have low return impedances back to their sources. For this to occur, the loop inductance in the layout of the power device must be kept *low*. Figure 6.14 gives two layouts (cross sections) for the drain and source of a power MOSFET on a printed circuit board. Note that the layout on the left-hand side has a larger current loop path from drain to source. This may be due to a layout designer for the board and a design engineer not paying careful attention to which power and ground planes they were connecting in the PCB layer stackup. It may also be due to layout constraints from signal density and routing. Here, the layout for the source pad has been connected to a plane on the opposite side of where the device is mounted. This results in the largest inductive or *current loop* path, and consequently, the largest inductance and, thus, potential noise. Conversely, the layout on the right shows a current loop path that is much lower, due to the ground plane being closer to the power plane of the device. Moreover, the engineer has chosen to attach the drain to a plane on the same side as the device, which resulted in a much shorter current

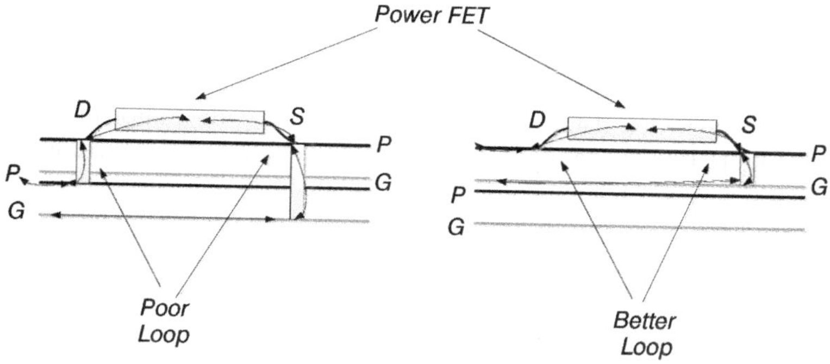

Figure 6.14 Comparison of two power FET layouts

loop. The question the engineer needs to ask is, What is the actual noise caused by the loop in either case, and will either result be a problem? Using the *first principles* equations from Chapter 3 to estimate these loops, it is possible to assess the impact of the potential noise on the system. This is illustrated in the next example.

Example 6.2 Estimate the loop inductance of the power FET in Figure 6.15. Use the dimensions in Table 6.4 to compute the loop inductance. Ignore any other loop paths outside of the package. Build a SPICE model using the schematic in Figure 6.13, and model the noise at the input. If the noise appears on the ground side as well, would you expect this noise to couple to the load? Ignore the PDN portion in the simulation.

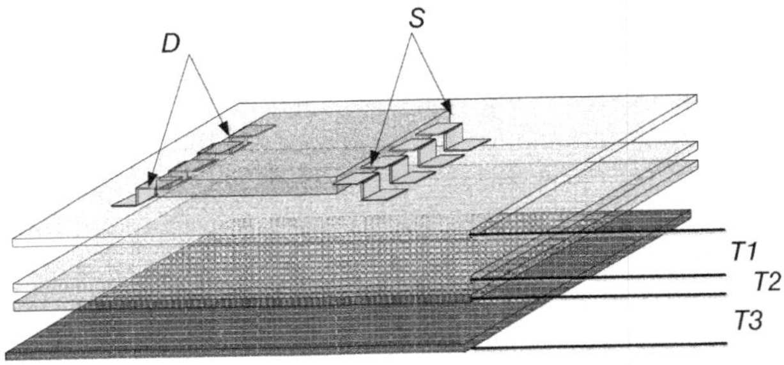

Figure 6.15 Layout of an FET package for noise estimation

TABLE 6.4 FET layout and PCB dimensions

Variable	Definition	Value
*T*1	Separation between *L*1 and *L*2	0.020 inch
*T*2	Separation between *L*2 and *L*3	0.010 inch
*T*3	Separation between *L*3 and *L*4	0.020 inch
PT	Pad pitch on package	0.030 inch
W	Pad to pad width	0.160 inch

SOLUTION: This is a simplified version of a complex problem. Ground loop coupling is often hard to detect and even harder to model. The illustration in Figure 6.15 is intended more to educate the reader about the source of the generation of this noise rather than suggest a methodology for the analysis.

The drain and source pin locations are shown in the package. The assumption is that the current flows from the drain to source and then loops back through the planes as shown in Figure 6.13. This *partial* inductance, essentially because the entire loop inductance will include the path through both bridges—from the filter back to the supply. The interest here is mostly in the supply loop inductance to determine how much noise this may generate. Here equation (3.111) in Chapter 3

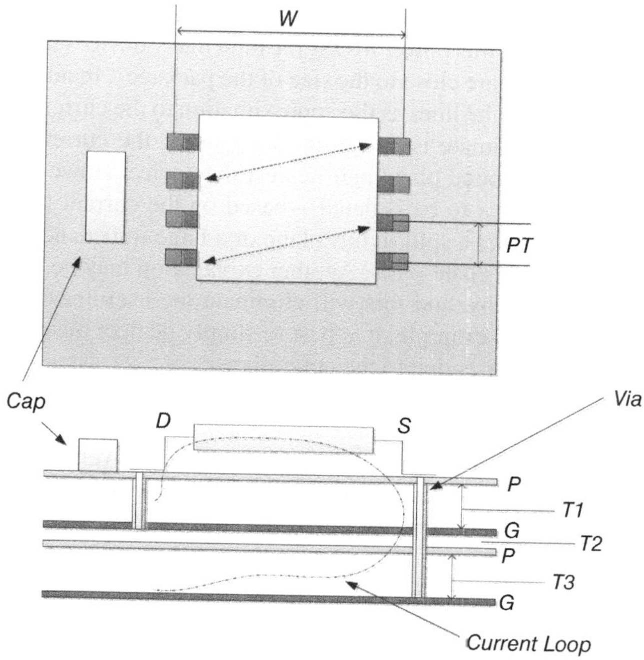

Figure 6.16 2D sketches for Example 6.2

can be used to compute the inductance below the FET for the two different sets of values.

To help in the estimation, it is a good idea to do a quick sketch of the structure in *two views*. Figure 6.16 shows the plan and cross section, with the dimensions carefully indicated. The main points of interests are the *average* values for the dimensions where the metal connections appear across a solder joint—or via joint in some cases—considered to be at the middle of the contact rather than at the inner or outer edge. The dimensions are given in a center-to-center arrangement to make the job a bit easier. This approach was used in Chapter 4 for a few similar computations. Once again, it is only an estimated average value for the inductance that is desired.

Because this is an estimate, the vias are lumped together rather than assigned the inductive contributions due to each. If the current flows in the directions shown in the top plan view of the figure, the width is used as the estimate of the current path through the planes. It can be assumed that the input capacitor is very close to the drain pads (as shown in Figure 6.16) so that any ancillary inductance outside of this loop can be neglected.

To a first approximation, the current should also flow through the planes of the board, mirroring that of the current flow within the package. However, it is understood that the silicon device and the transistors are not spread out across the package width. This information about the internals of the silicon is difficult to get. Nonetheless, it can be assumed for this power MOSFET (and many devices used today) that the transistors and silicon are close to the size of the package.[4] In addition, one can take the distance between the lines as the approximation to the current flow through the package. The last estimate is due to the loop itself, the current path will be bounded by the power ground plane pair nearest the device. However, notice that the middle P/G pair appear to be isolated—based on the current flow path. This indicates that there is either a split in this plane or a large impedance between this plane-pair and the outside plane-pairs. Another explanation may be that this region has signals running near by, and this will eliminate the usefulness of the planes underneath them. For this example, it is best to simply neglect this possibility and assume that the inductance is due to the outer planes.

Neglecting the thickness of the planes, and with a clear understanding of the dimensions and current flow in the sketch, the approximate data needed to plug into the equation are now available. The dimensions were given in inches, so it is necessary to convert them to meters, since the variables in the equation are metric based. Using equation (3.111) the result becomes

$$L = \frac{\mu_0 ls}{w} = \frac{\mu_0 W(T1 + T2 + T3)}{2PT} = \frac{(4\pi e^{-7})(4.06e^{-3})(1.27e^{-3})}{2(1.52e^{-3})} \cong 2.13\text{ nH} \tag{6.9}$$

The next task is to simulate the noise in SPICE. The ground return is placed on the side where the input capacitor ground will be located, and then the inductors are placed on the drain side of the upper FET and on the source side of the lower FET. The grounds are removed everywhere else so that the noise can be examined. Note in Figure 6.17 that there is significant noise at both the ground and power sides. If a finite inductance exists from this point to the output of the PDN, it can be assumed that there is a high likelihood that this noise will couple to the load side as well.

As was stated previously, a number of factors need to be considered in order to determine the exact amount of noise at a load, and among these factors is the impedance of the ground return path from the silicon device. If the noise is purely *differential* in nature, it will *appear* that the noise does not affect the guardband. The noise will nevertheless have an effect on the voltage guardband because of the difference in ground references between the VR controller and the load. The noise impacts from the VR will be discussed some more later in the chapter.

6.2.2 Coupled Power Bus Noise

Coupled power bus noise occurs when two power delivery networks overlap in some manner, allowing the noise to capacitive and inductively couple between them [16]. Where this noise is most prevalent is when relatively large power planes are

[4]This is not always the case for older devices. Newer devices have smaller packages where the bodies of the devices are only slightly larger than the silicon that is housed within.

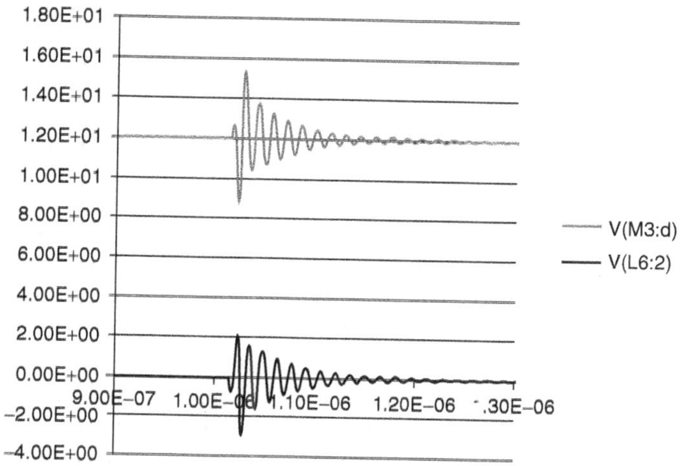

Figure 6.17 Noise on power and ground sides of the model

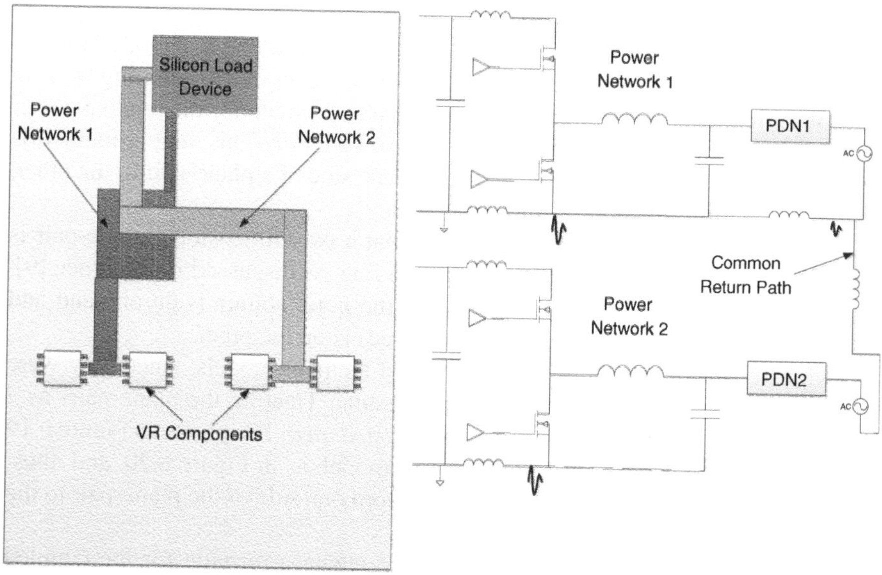

Figure 6.18 Coupled power bus noise example

vertically stacked close to each another [17]. This may occur in PCB stackups where the cost and layer counts are at a premium and where constraints force the board designer to allow voltage planes to coincide with one another. Figure 6.18 shows an example of a structure where noise may couple in from one network to another. One network is considered to be the *susceptor* (PDN 1) and the other the *aggressor* (PDN2).

If the return paths are separated or have some impedance between them, this may help mitigate (potentially) some of this noise. However, in most practical systems, the ground structure is *shared* (as shown in the figure), providing the opportunity for noise to couple between the networks. The result of this coupling, with respect to the voltage guardband, is the same as that noted in Section 6.1.1, a *reduction* in the guardband. The question is, how much? To answer this, requires the PI engineer to develop a working model of the problem and then simulate to the limits. As in Example 6.2, the problem (again) is quite complex, and even efforts to model it adequately can be challenging due to the inability to accurately model all of the components that can contribute to the noise in the system. Besides, the many factors that can cause the noise to couple make it very difficult to isolate the main forcing function. Fortunately, for the voltage guardband, the goal here is to determine an *upper bound* to the problem and then assess if the impact is significant. Once it is determined what the impact to the voltage guardband is, a number is assigned to the overall guardband—where that number is eventually given to various groups to determine the performance, binning, and reliability of the silicon as it operates within the system.[5]

A simple method, developed by Hu et al. [9] uses a transmission line model to estimate the noise at resonance. This method does not predict the coupling between planes that overlap—this will be addressed shortly. Moreover, the model appears to be mostly intended for determining EMI noise on planes due to coupling from an adjacent source. However, it does give a reasonable estimate for the relative power bus noise from the noise *generator* to the noise *receptor*. This simple method is very useful for predicting the peak noise from one side of a plane-pair to the other, and it is worth repeating and illustrating here.

One starts by assuming, for the moment, that a two-dimensional plane-pair is represented as a one-dimensional transmission line, as discussed in reference [9]. Figure 6.19 shows a power plane-pair where the noise source is on one end and the measurement point (where noise is measured) is on the other.

Previous chapters mostly addressed lumped element models, since they were simple and easy to compute for the desired results. Treating the plane-pairs as a transmission line assumes that the structure is *distributed*. The planes in Figure 6.19 can now be represented as a transmission line model, as in Figure 6.20, and, thus, in this simple way used to compute the noise from one side of the plane-pair to the other.

From transmission line theory, it is easy to create a formula for the coupled S-parameter (S_{21}) between the generator and the receptor. An *S-parameter* is essentially a transfer function in the frequency domain in either a one- or two-port network. In this case, the S_{21}-parameter is the gain from the ratio of output to input voltage, or

$$S_{21} = \frac{V_{Out}}{V_{In}} \tag{6.10}$$

[5]This information is also key to the board and platform teams who must deliver a system that operates within the boundaries and specifications defined by the design and marketing teams and others.

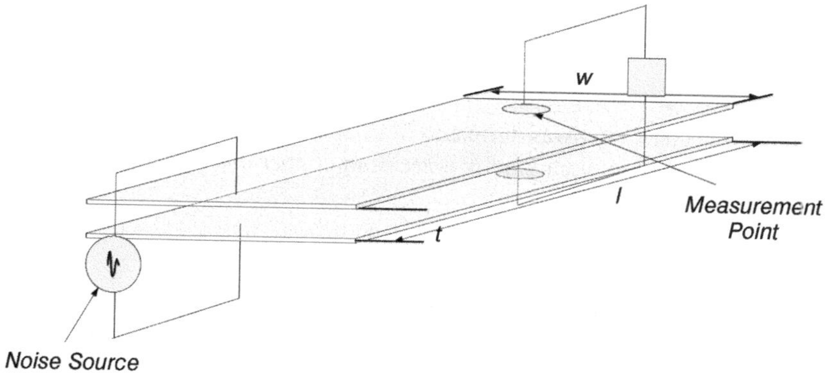

Figure 6.19 Representation of power plane-pair for transmission line estimation

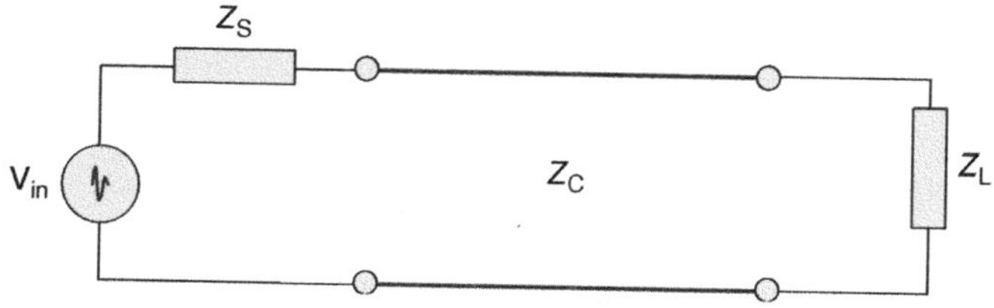

Figure 6.20 Transmission line model for power distribution bus

The model assumes that the transmission line is *lossy*, though the model may also be used for lossless systems. The equation for the S_{21}-parameter between the two points in the planar structure is

$$S_{21} = \frac{2}{e^{\gamma l} + e^{-\gamma l} + \left(\frac{e^{\gamma l} - e^{-\gamma l}}{2}\right)\left(\frac{Z_C}{Z_0} + \frac{Z_0}{Z_C}\right)} \tag{6.11}$$

where, as in Figure 6.20, Z_C is the characteristic impedance of the PDN (in this case, just the planes), Z_L is the impedance at the load, and Z_S is the source impedance of the noise voltage source. The impedance Z_0, here, is the impedance of the measurement device in the system—it is assumed that this is 50 Ω, as is common for many measurement devices such as a vector network analyzer. Most of the parameters in equation (6.11) were encountered in Chapter 3 in the discussion of the wave equation. The parameter γ is related to the propagation constant (β as it is sometimes defined)

$$\gamma = \alpha + j\beta \tag{6.12}$$

β is related to the wavelength

$$\beta = \frac{2\pi}{\lambda} \tag{6.13}$$

The attenuation constant α was discussed in Chapter 3 in terms of copper losses. If the frequency is high enough, one will see both copper *and* dielectric losses, and thus α is related to both

$$\alpha = \alpha_c + \alpha_d \tag{6.14}$$

For the analysis here, it will be assumed that the dielectric losses are negligible, and thus, only the copper losses will be considered. The copper losses is determined using

$$\alpha_c = \frac{R_S}{wZ_C} \tag{6.15}$$

where R_S is the *surface* resistance of the plane. The surface resistance is determined from knowledge of the skin depth of a material [19]:

$$R_S = \sqrt{\frac{\omega\mu}{2\sigma}} \tag{6.16}$$

which actually follows directly from the surface impedance for the structure—in this case the surface resistance is the *real* part of the equation.

Hu et al. [9] points out that if the peak values at resonance are of interest, then equation (6.11) is simplified to the following relation:

$$S_{21} = \frac{2}{2 + \left(\frac{Z_0}{Z_C}\right)\alpha_C l} \tag{6.17}$$

For frequencies below 1 GHz, and for relatively small PCB planes (relative to the wavelength), it can be seen that the relation in equation (6.17) yields reasonable approximations. The S_{21} value can now be converted to determine the voltage noise coupling from the source V_S—as was our initial goal. The next example illustrates the use of this method.

Example 6.3 Estimate the coupled noise in a system using the transmission line method outlined above. Assume the parameters for the plane structures listed in Table 6.5. Use Figure 6.19 for the dimensions, and assume that the measurement point is approximately the length of the plane-pair. Plot the S_{21}-parameter as a function of frequency.

SOLUTION: The first step is to compute the characteristic impedance of the structure, using equation (3.111) again from Chapter 3 for the inductance. Recall,

TABLE 6.5 Parameters for S21 computation

Variable	Value
w	10 mm
t	50 μm
l	20 mm
ε_r	4.2
Z_0	50

from transmission line theory, that the characteristic impedance is computed from knowledge of the inductance and capacitance of the structure. Fortunately, the *first principles* formulas give reasonable estimates for this once the geometry of the structure is known. The result will be a *lumped* value. However, again, recall that for a transmission line, the impedance is determined by the per-unit-length parameters of the structure [18]:

$$Z_C = \sqrt{\frac{L}{C}} \tag{6.18}$$

The inductance and capacitance are computed from the dimensions of the structure. This way the impedance can be computed directly. Since the same inductance formula was used in Example 6.2, it is only necessary to show the results for the inductance and thus compute the capacitance below:

$$C = \frac{\epsilon_r \epsilon_0 A}{d} = \frac{(4.2)(8.854e^{-12})(20e^{-3})(10e^{-3})}{(0.05e^{-3})} \cong 300\ \text{pF} \tag{6.19}$$

The inductance, L, is around 0.63 nH. Thus, the characteristic impedance is,

$$Z_C = \sqrt{\frac{L}{C}} = \sqrt{\frac{64e^{-12}}{300e^{-12}}} \cong 0.46\ \Omega \tag{6.20}$$

Now, substituting α_C and R_S into equation (6.17) obtains

$$S_{21} = \frac{2}{2 + \left(\frac{Z_0}{Z_C}\right)\alpha_C l} = \frac{2}{2 + \left(\frac{Z_0}{Z_C^2}\right) R_S l w} = \frac{2}{2 + \left(\frac{Z_0}{Z_C^2}\right)\sqrt{\frac{\omega\mu}{2\sigma}}\, l w} \tag{6.21}$$

The resulting function can then be plotted over the frequency range of interest to determine the magnitude of the noise at each frequency.

As noted by Hu et al. [9], equation (6.21) gives a simple formula that shows the envelope for the S_{21} parameter over the frequency range of interest. Figure 6.21 plots this ratio from 1 MHz to 1 GHz. The noise coupling is quite large at the lower frequencies (as expected), and thus, one needs to ensure that the PDN will filter this

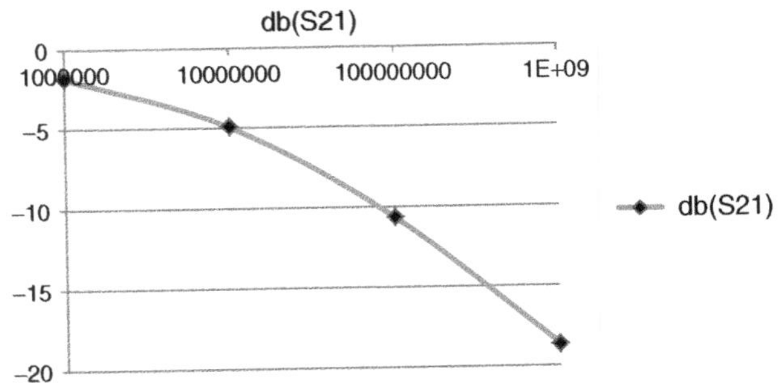

Figure 6.21 Plot of S_{21} in decibels for Example 6.2

noise with good quality capacitance in this frequency range and that the noise does not impact the guardband margins. Clearly, as is apparent in the figure, the noise *does* reduce due to the attenuation in the planes as the frequencies increase—and this indicates that the voltage guardband is less of a problem above a certain frequency. By definition, the voltage guardband is a *low frequency* to DC issue, so the noise for the guardband would be addressed when the frequency is relatively low. This example is instructive because it illustrates that noise can couple from one side of a PDN to the other, and without the addition of filtering and the capacity to reduce the guardband, significant noise may couple into the load or other parts of the system.

Observing how noise can couple onto a plane-pair, and analyzing how noise couples from one pair to another is the next step. Feng et al. [16] utilized the *segmentation method* to show how noise couples between planes that overlap, as shown in Figure 6.18. This method also uses a *cavity* model to create analytical expressions for the impedances that map into a Z matrix for computation. Okoshi [20] proposed a segmentation method as a way to analyze microwave networks for power distribution. Feng utilized this method with the intention of simplifying the analysis of complex plane geometries to aid the designer in the analysis of noise coupling in multi-power plane structures. This method is now discussed with respect to power plane coupling.

Figure 6.22 shows a plan and side view of a structure where two power plane-pairs overlap. The objective is to analyze these plane-pairs as simple individual structures with ports. The ports connecting them allow the impedances of each to be isolated. With the impedances isolated, they can be analyzed independently by using a cavity model (or another model depending on the required accuracy) and then solving the final matrix for the voltages.

In Figure 6.22, the left-hand side shows the power ground plane-pairs with external ports *A* and *B*—in both a plan view and a cross-sectional view. It is assumed for this analysis that the planes completely overlap. This is not necessary for the method to work as is evident in the references. The planes are essentially *folded out* and then re-connected with internal ports *C* and *D*, as shown in the figure on the

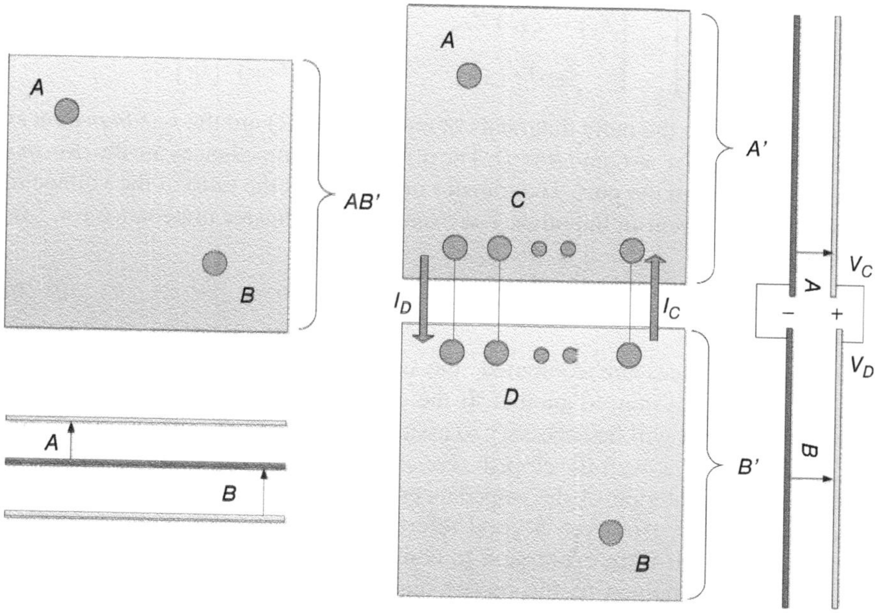

Figure 6.22 Power plane-pairs illustrating the segmentation method

right-hand side. The two plane-pair segments A' and B' have voltages and currents that go between them. As can be seen in the figure, there are two patches, A', B', and AB'. Next, assign Z-matrices to each patch as $Z_{A'}$, $Z_{B'}$, and $Z_{AB'}$. As noted by Feng, it is possible to compute the Z-matrices in a number of ways. The composite matrix $Z_{AB'}$ can be computed by applying the interconnect between the ports. The new matrices, $Z_{A'}$ and $Z_{B'}$, are broken into submatrices that are made up of both internal and external port connections:

$$Z_{A'} = \begin{bmatrix} Z_{AA} & Z_{AC} \\ Z_{CA} & Z_{CC} \end{bmatrix}, \quad Z_{B'} = \begin{bmatrix} Z_{BB} & Z_{BD} \\ Z_{DB} & Z_{DD} \end{bmatrix} \tag{6.22}$$

The voltage and currents at the external and internal ports are

$$\begin{bmatrix} V_A \\ V_C \end{bmatrix} = Z_{A'} \begin{bmatrix} I_A \\ I_C \end{bmatrix}, \quad \begin{bmatrix} V_B \\ V_D \end{bmatrix} = Z_{B'} \begin{bmatrix} I_B \\ I_D \end{bmatrix} \tag{6.23}$$

Note that the currents and voltages at the internal ports have the following relations:

$$V_C = V_D, \quad I_C = -I_D \tag{6.24}$$

Then substitute in for the voltages and currents and the submatrices to get the following matrix equation:

$$\begin{bmatrix} V_A \\ V_B \end{bmatrix} = \begin{bmatrix} Z_{AA} - Z_{AC}YZ_{CA} & Z_{AC}YZ_{DB} \\ Z_{BD}YZ_{CA} & Z_{BB} - Z_{BD}YZ_{DB} \end{bmatrix} \begin{bmatrix} I_A \\ I_B \end{bmatrix} \tag{6.25}$$

The impedances on the main diagonals of equation (6.22) are the *self-impedances* between the ports if $A = C$ and $B = D$. The off-diagonal impedances are the *transfer impedances* between the port. The transfer impedance is the ratio of the voltage in one port to the current in the other. For example, the transfer impedance for, say, Z_{AC} results in

$$Z_{AC} = \frac{V_A}{I_C} \tag{6.26}$$

Once the impedances are computed, the engineer can determine the voltages from one coupled system to another. If the cavity model is used to resolve these impedances, then it will be necessary to assume that the frequency is high enough such that there is a well-developed *skin-effect* in the planes (as discussed in Chapter 3). Computation of the impedances via the cavity model is beyond the scope of this book. However, a good reference for the reader is given in [21], which uses a closed-form expression to compute these impedances. As with any approach, the limitations of the model must be considered to ensure that the approach is applicable to the problem at hand. Moreover, assumptions can be made for the frequency range of interest that ensure a reasonably accurate result.[6]

6.2.3 Simultaneous Switching Noise

Simultaneous switching noise, or *SSN*, is an important study in many disciplines with respect to electromagnetic behavior at the circuit board, power bus, and silicon level. The subject arises extensively in discussions of cross-talk and noise within signal integrity problems and can be an important concern at both the receiver and driver for a high performance IO structures either on chip or within a silicon device [22]. SSN is typically caused by large perturbations at the silicon device, or load, that result in ground bounce across an inductive interconnect from the switching devices, simultaneously causing relatively large currents that inevitably show up as a high-frequency voltage shift off of the main ground voltage local to that device. Figure 6.23 shows this electrical bounce more clearly than words can. In the figure, a set of devices is switching at the same moment, or at virtually the same time, such that their currents add up *simultaneously* at the same electrical junction. The currents funnel through the lowest impedance path that is local to these devices, and a relatively large instantaneous voltage shift occurs at that point. The voltage drop causes the electrical signal to move relative to the ground potential, causing the timing distortion. The result of this shift is that signal fidelity is compromised and thus the potential for data corruption exists.

[6] It is critical to be wary of the lumped element assumptions here if the PI engineer is using an L-C circuit model. This is because the lumped model does not take into account frequency dependence. In this case, the engineer should use an approach that computes the matrices if a higher level of accuracy is warranted.

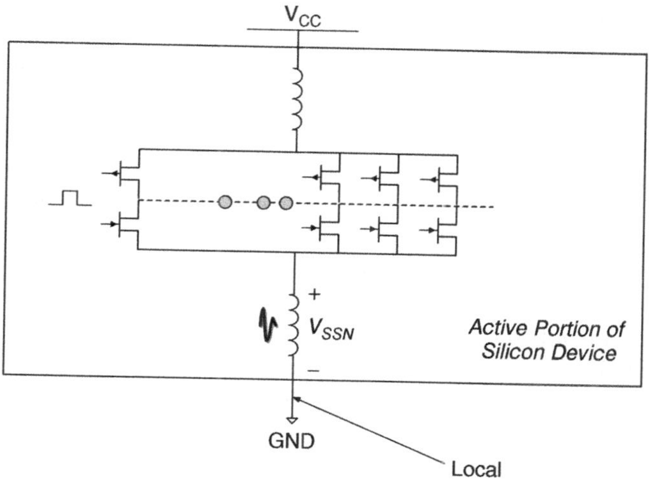

Figure 6.23 Illustration of SSN effects

The power integrity issue is related to that of signal integrity, since the ground bounce of an SSN voltage shift can affect the quality of the power delivery to the silicon load. Yet, this problem is more than a *high-frequency* issue, because the ground bounce noise is typically considered a *local* phenomenon, it is filtered out (due to the very high frequency nature of the currents) before it ever reaches the voltage regulator and most of the PDN network. The SSN normally occurs during the switching *edge* of the signal, which is typically in the 100s of pico-seconds—or, in the frequency domain—in the 5 to 10 GHz range—which is definitely out of the bandwidth of most PDNs. After the shift in the ground occurs multiple times over multiple cycles, it could appear to the regulator as a normal ground shift. This is how: the remote sense lines of the regulator are tied to both the ground and voltage planes nearest the load. There is a filter, on these sense lines, that screens out the high-frequency noise. However, the local ground and the global ground are no longer the same. What may cause this issue is a *return path discontinuity* (RPD) in either the voltage plane reference or ground plane reference for a set of signals [22]. An RPD can also occur in the PDN return path. The result is an impedance that can force a voltage differential between the VR source and the silicon load. This offset results in a misrepresentation of the voltage and ground reference relative to the logic, and the delivered voltage is lower or higher than expected. The problem with SSN for the PI engineer working at the platform level is that there is little that can be done to mitigate this noise—it is almost entirely a local silicon problem—and often the PI engineer, and the package and system silicon integrator, have separate responsibilities that can disconnect the PI engineer from the SSN problem. This is why when noise and/or performance issues are discovered in the lab, the engineers should come together and find the cause of the problem.

6.3 POWER NOISE REDUCTION TECHNIQUES

So far this chapter has discussed how noise is generated and how it can affect voltage guardband and other aspects to the power delivery quality. We are ready now to consider how to mitigate the noise problems. When noise couples into a PDN there are two options: fix the fundamental cause of the noise or reduce the noise with a fix that reduces it to an acceptable level—the latter is considered a *Band-Aid* to the problem. This is a classic dilemma for a team that is responsible for delivering a high-quality solution. Which option is chosen is a function of time, resources, and cost—and, of course, the efficacy of the solution. Thus, this discussion will first investigate common types of noises and then attempt with some basic techniques to reduce the noise. For the PI engineer, there are certain limitations as to what can be done, since often noise tends to be generated and coupled in regions that are outside their scope and control. Sometimes, though, the problem may not be within the PI engineer's jurisdiction, but the solution may be.

The first step is to examine where the noise is present and what the problem is. Recall from Section 6.2 that there are a number of possible noise sources in a system. The usual source of noise is the voltage regulator. The VR switching noise is mitigated through good layout techniques (as shown in Section 6.2.1) or through better VR design (e.g., a topology change with a soft switch design). Another way to reduce VR noise is to add decoupling to the VR. Example 6.2 illustrated this point by a capacitor placed very close to the switch. Another consideration may be to change the frequency content of the switching noise by slowing the ramp rate of the switch as it turns on. While this approach could increase loss in the VR, it is useful for many systems where this trade-off is acceptable. For the PI engineer, sometimes getting every option on the table is necessary. Last, it may be possible to reduce noise by adding capacitance to the PDN at a certain point in the distribution path. Figure 6.24 shows this approach. The addition of HF capacitance near the power converter side of the PDN *squelches* some of the noise coupling into the PDN.

As discussed in Chapter 4, the PDN is a distributed network that can be broken down into low-frequency bulk, mid-frequency, and high-frequency capacitances. Typically, the high-frequency capacitance is placed nearest the load because this is where the need is most. However, it is can *also* be placed near the VR to reject the noise generated via the switches from the VR. The effect of adding this capacitance can be seen on the PDN frequency response to the impedance in Figures 6.25 and 6.26.

In Figure 6.25, the observed resonance is in the 3.5 to 85 MHz range. The first resonance would be close to the switching frequency of a converter that switches in the 1 to 3 MHz range. The second is within the bandwidth of the edge-rate of the on/off rise and fall time of the FET switch. It is a simple matter to determine the approximate frequency content of this edge-rate as [23]

$$\omega = \frac{\pi}{t_r}, \quad \text{or} \quad f = \frac{1}{t_r} \tag{6.27}$$

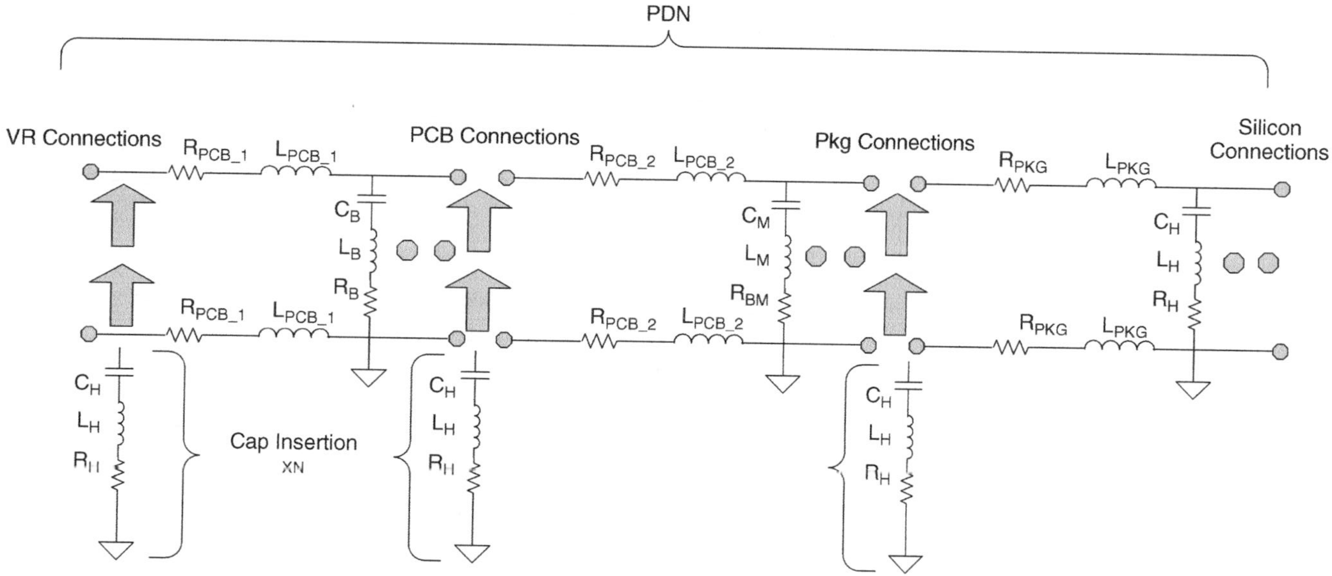

Figure 6.24 PDN distribution with high-frequency caps inserted in LF areas

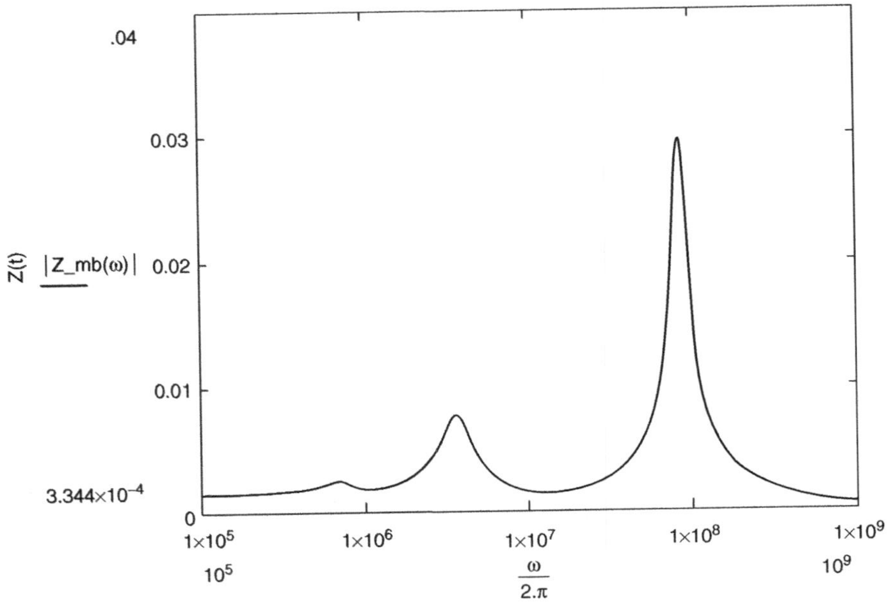

Figure 6.25 Mathcad plot of PDN impedance without HF caps

This equation comes directly from the Fourier series of a trapezoidal pulse where, as was seen in Chapter 5, the result is a *sinc* function. The rise-time, defined here as ranging from 0 to 100%, is equivalent to the center frequency of the sinc function after taking the Fourier series of the time-domain trapezoidal pulse. If the power FET is assumed to be able to turn on quickly, and align with the resonance shown in Figure 6.25, then the frequency in the 50 to 100 MHz range would put the rise-time in the 10s of nanoseconds. Also any ringing noise will be close to this frequency. There are two ways then to reduce the noise at the resonances; first, reduce the resonances in the PDN so that the noise voltage is *attenuated* at these points, and second, shift the frequencies of the resonances so that they do not align with the VR switching—for example, the frequency of the FET switching does not align with the resonances of the impedance of the PDN. In Figure 6.26, it can be seen that the resonances have been reduced and shifted in frequency from the resonances in Figure 6.25.

The capacitors added at the nodes, as shown in Figure 6.24, have effectively accomplished our objectives.[7] But note that only *high-frequency* capacitance was added—the difference here is in the *quality* of the capacitance. The *parasitics* of the added capacitors are significantly smaller, which effectively reduces these peaks and shifts them. The peak at 85 MHz was shifted upward to around 150 MHz and then attenuated by approximately half the value. The 3 MHz resonance was also

[7] The additional capacitance was essentially a few high quality MLCCs (~5) in each region, which had relatively low ESLs. The capacitance values were not important but the inductance was.

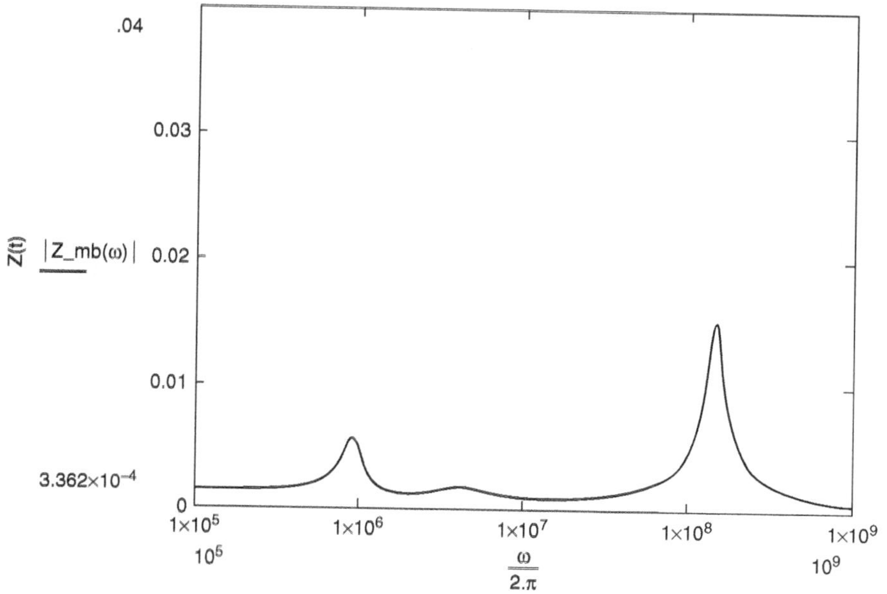

Figure 6.26 Mathcad plot of PDN after adding HF caps

reduced and shifted to approximately 6 MHz. If the generated noise occurred at 70 MHz or so, the noise would have to be reduced by at least a factor of 4× by adding a group of high-frequency capacitors in the PDN path. Another important point to note is that the capacitance nearest the VR to PDN nodes create a *shunt* path back to the return path, which will also steer the ringing currents back to the source, reducing the coupling into the PDN where the silicon resides.

Though this may not be the most effective way to reduce the noise, may be the most *cost*-effective overall—assuming that adding some capacitors to a PCB is easier than changing the VR design—and is within the scope of the PI engineer's job description.

6.4 EMI CONSIDERATIONS FOR POWER INTEGRITY

Though EMC (electromagnetic compatibility) is not in the immediate regime (typically) for the power integrity engineer, there are still aspects to compliance that need to be considered. Virtually all computer systems sold are required to meet certain electromagnetic emissions requirements (e.g., FCC Class A or B), and thus, the generation of noise at the chassis, board, and silicon is something that affects all designers in one way or another. With respect to power integrity, it is important that the PI engineer understand how the design, layout, and interaction of the PDN and power delivery maintain, or rather, do not *disrupt* the certification process for the platform.

Most small computer systems, such as Laptops, are required to meet FCC Class B regulations for domestic environments. For larger systems such as servers, they must meet Class A requirements. Some products intended for nonresidential/domestic environments may also be used for residential—but, because these products may emit more noise, the user may need to take care that the interference is low enough that it does not interrupt other devices. The requirements are given in *decibels* of noise as measured in a chamber or an *OATS* (open area test site). The unit of measure is either the *electric* or *magnetic* field. Thus, for the electric field, the relative emitted noise measured will be a ratio. This ratio is relative to a given noise floor, or for the FCC standards, 1 μV. Thus, if one mV of noise were measured from a radiating source, the amount of noise in decibels would be

$$\text{Decibels of noise} = 20\ \log\left(\frac{1e^{-3}\ \text{V/m}}{1e^{-6}\ \text{V/m}}\right) = 60\ \text{dB} \tag{6.28}$$

EMC engineers make their measurements in decibels because most of the regulations are defined in these terms.

There are two types of EMC emissions at the system level; *radiated* and *conducted*. Paul [18] discusses both types, and it is recommended that the reader review this reference for an excellent overview. Conducted emissions are simpler than radiated, in that the measurement is made at the power cord and is done by placing a current probe around the cord to pick up the conducted noise at this point. Figure 6.27 shows how noise may move from the circuit board and chassis, out through the power cord. The noise may start from any active device where either *common-mode* or *differential-mode* currents are generated. Common-mode currents travel along a path in the same direction away from (typically) the source. This path may consist of a power/ground plane-pair or simply two parallel conductors. Differential-mode currents travel in equal and opposite directions along the

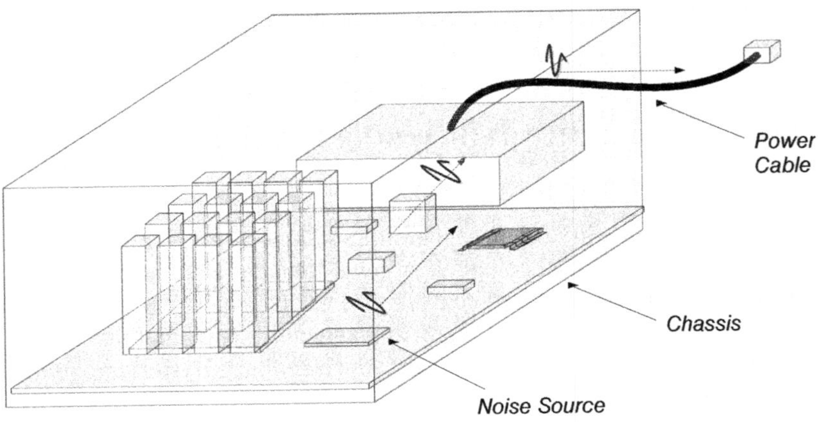

Figure 6.27 Conducted noise path in a chassis

conductors. The main difference between these currents, with respect to EMI, is that common-mode currents generate more emissions and are often harder to limit in a system. Both currents are generated from some type of *source*—the source can be a device on a printed circuit board that switches periodically or a-periodically. Most offenders for the source of noise are high-frequency clocks, high-speed IOs, high-performance computing devices, and power converters. If the ground loops are not well designed for the device, switching noise may find its way across the printed circuit board and into the power supply. Then, if proper filtering was not employed, the noise may find its way onto the power cord and thus emit into space. For conducted emissions, however, the frequency limits are fairly low (30 MHz), and thus, beyond this frequency the more critical emissions will be of the radiated type. In some cases, it is required to add a filter (e.g., a common-mode choke) to reduce these emissions to an acceptable level. The filter absorbs the conducted energy the frequencies of interest, which reduces the electromagnetic noise to an acceptable level.

Radiated emissions are more complex because there are many different ways in which noise can radiate out of the chassis and off the PCB. Moreover, the bandwidth for the regulations is higher than that for conducted emission regulations. Today, because of possible interferrence with RF communications, EMI must be kept to a minimum in many systems, in order to ensure the integrity of the radio waves (and therefore, data) being transmitted through space. The measurements for such noise is typically done in a chamber where the DUT (device under test) or source is located in one part of the chamber and the antenna, or receptor, is located a given distance away (depending on the chamber this could be 1 to 3 meters). The chamber is shielded from outside noise sources and is calibrated to a given noise floor, as previously defined. The measurement equipment is typically a type of spectrum analyzer (SA) that measures the emitted noise through a given limited bandwidth window. Figure 6.28 shows how a typical measurement is made in a chamber to determine the emissions from a particular chassis/DUT. Depending on whether the noise is common mode or differential mode, the noise limits within a given bandwidth will be different.

Regulations for FCC Class A and B at 50 MHz are given in Table 6.6 and Table 6.7 [27]. The fields radiating from a DUT are complex and the electric and magnetic fields that radiate are *spherical* in form. Though the EMC engineer will measure the results directly from a closed chassis, measurements can also be made off the PCB in an open environment. Often, the EMC engineer will probe parts of the board to determine where the noise is coming from and what the cause is. The radiated noise in the far-field rolls of as $1/r$, where r is the radius of the field to the measurement point from the point of origin—in this case the DUT or board. Near-field amplitudes typically roll off at $1/r^2$ and $1/r^3$ for magnetic and electric fields, respectively. The distinction between the near and far field for a *dipole*, for

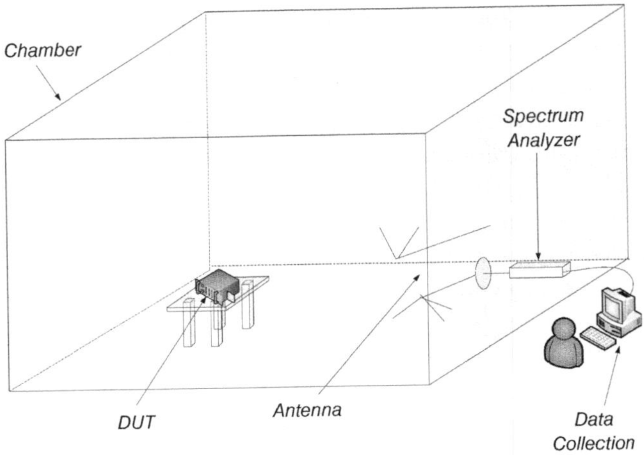

Figure 6.28 Example of radiated emissions measurement setup

TABLE 6.6 FCC class limits, Class A and B at 50 MHz

Regulation	Limit	Distance
FCC Class A	90 μV/m	10 m
FCC Class B	100 μV/m	3 m

TABLE 6.7 Typical PDN related issues causing potential EMI noise

Noise-Generating Sources
Signal vias through a power bus
Power plane discontinuities
Ground or power plane inductive routes
Poor placement of decoupling capacitors or poor choice
Poor routing of power/ground connections through connector or socket with silicon
Poor choice of local and/or global system ground point relative to PDN ground

example, comes when the wavelength of the field is 1/6 that of the radius, or

$$r = \frac{\lambda}{2\pi} \cong \frac{1}{6}\lambda \tag{6.29}$$

Additionally, a crude way of estimating the noise of a board is to represent the source as an *electric dipole*. As shown in Figure 6.29, the dipole resides on the z-axis.

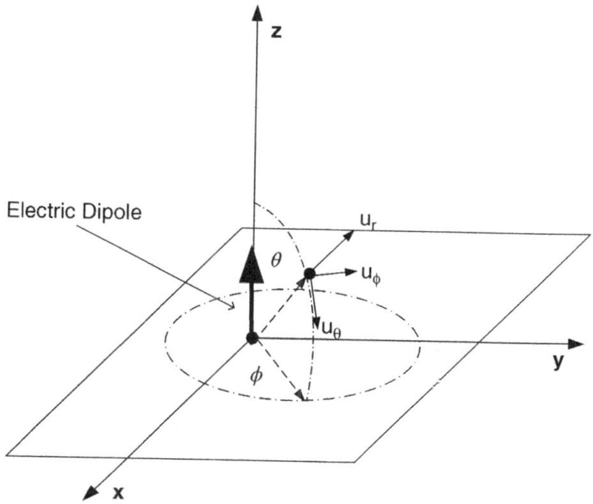

Figure 6.29 Electric dipole in spherical coordinates

The dipole is used to represent a source-carrying current on a PCB, such as a power plane with a current. Since the interest here is in relatively high frequencies with respect to radiated emissions, the discussion will be limited to 200 MHz and above. Below this frequency, EMI engineers can rely on good chassis design to shield noise due to the larger wavelengths.

The electric and magnetic fields for an electric dipole, in the far field, are shown below for reference:

$$\vec{E}_{FF} = j\eta_0\beta_0\frac{\hat{I}l}{4\pi}\sin\theta\,\frac{e^{j\beta_0 r}}{r}\vec{u}_\theta \tag{6.30}$$

$$\vec{H}_{FF} = j\beta_0\frac{\hat{I}l}{4\pi}\sin\theta\,\frac{e^{j\beta_0 r}}{r}\vec{u}_\phi \tag{6.31}$$

The *magnetic* dipole, or *current loop* equations, are found in [18]. Here, the terms in equation (6.31) and (6.32) are as follows: η_0 is the intrinsic impedance of free space, the current vector I is the current at the source of the dipole in the z direction, and l is the length of the dipole. Note the vector directions of the fields in the far field—where the electric and magnetic fields are orthogonal to each other (as expected). The fields are essentially *plane waves* propagating in air at this point. To use these equations to approximate the noise from a PCB, the results are simplified. For example, if one has a power/ground plane-pair on the surface of a board, pointing in the z direction, where the currents are strictly differential mode, the magnitude of the electric field will be

$$|\vec{E}_{FF}| \cong 1.316 \times 10^{-14}\,\frac{|\hat{I}|f^2 ls}{d} \tag{6.32}$$

where s is the separation of the planes (center to center) and d is the distance from this center to the measuring point (e.g., the antenna). This is the result as given by Paul [18]. As another example, if one has a PDN strip of 5 cm on the surface of a PCB with a differential mode current of 30 mA at 400 MHz, separated by a distance of 5 mm, in a 3 m chamber, one would expect the electric field strength to be

$$|\vec{E}_{FF}| \cong 1.316 \times 10^{-14} \frac{(30e^{-3})^2(400e^6)^2(5e^{-2})(5e^{-3})}{3m} \cong 160\ \mu\text{V} \qquad (6.33)$$

or around 44 dB. Note that some assumptions must be made to ensure this is a *reasonable* result, namely that the width of the planes must be small relative to the other dimensions (s and d) and that the currents at the frequency of interest are uniform—which, in most cases, is a reasonable assumption. In some cases, as discussed previously, the EMC engineer may make near-field measurements for the PI engineer to determine where the noise is coming from. These measurements, of course, are possible, provided that the board design is somewhat finished and that the *final* design is now *after the fact*. Sometimes the measurements are made at the request of the PI engineer, to get an understanding of whether or not the PDN layout and design created excessive emissions. Though the PI engineers' PDN may not be the source of the noise that is generated in a system, it is crucial that the decisions made with respect to the power distribution layout and overall system power delivery not limit the EMC engineer from making sure the system is certified. This means that not only must the PI engineer implement a satisfactory PDN for the system but also must not create additional noise, or allow this noise to propagate unnecessarily across the planes and traces of the PCB.

There are a number of mechanisms in the design of the planes and the PDN that can lead to the generation of emissions. To discuss the details on each one would be beyond the scope here, but a brief description of some of the more common mechanisms is warranted. Many such problems appear to be simply a matter of *common sense* for the designers involved, but when layout and system constraints come to bear, basic noise mitigation strategies can take a back seat to more seemingly pressing matters. Thus, it is a good idea to have a level of understanding of what can happen if the engineer is not vigilant. Some of the causes of noise are listed in Table 6.7.

Signal vias through a Power Bus Often, signals must be routed through a quiet power bus because of routing constraints. However, this presents an opportunity to couple noise into the plane—particularly if the signals are noisy, such as high-speed IOs or clocks [28]. This is particularly true if the same signal serpentines through the plane from one layer to the next stitching its way from one side to the other.

Power Plane Discontinuities This is similar to the signal via issues above but is more general. This is mainly around the power plane distribution having discontinuities in it—such as abrupt necking down in regions, capacitor bridging across planes, and large regions where the areas are void of capacitors along the PDN. The main problem here is that discontinuities can generate common-mode currents, which can generate significant EMI in a system.

Ground and/or Power Plane Inductive Routes Another problem arises if the power plane must be routed through multiple regions, like a signal trace. This results in potentially large inductances that can resonate with signal noise if the noise couples into them. As usual, it is best to try to minimize this layout problem to help mitigate system level EMI.

Poor Placement of Decoupling Capacitors or Poor Choice Depending on the frequency of noise in the system, poor placement of capacitors (e.g., large inductive connections to the plane or placement relatively far away from the power/ground planes) can accentuate noise issues as well. This is quite common, since often the layout engineer and PI engineer may not be fully engaged at the proper time (or sometimes not at all). Moreover, it is easy to put down the wrong type of capacitor for a section of the distribution network if specific instructions are not given to the layout engineer.

Poor Routing of Power/Ground Connections through Connector or Socket with Silicon This problem occurs mostly when the package designer does not have enough power/ground connections in the package or the placement of those connections is not optimized due to other constraints. The result is discontinuities in the routing, which in turn causes noise coupling into the planes and thus generation of platform level EMI noise.

Poor Choice of Local and/or Global System Ground Point Relative to PDN Ground There are two schools of thought on grounding a PCB to a chassis: local grounding at a single point, or global grounding throughout. For the PI engineer, it is critical that the PDN grounds be local to the silicon and power converter in order to keep ground loops contained close to their sources. When the board is grounded—whether it be local or global—the PI engineer should make sure that their PDN ground reference has a low impedance *local* loop and does not have a significantly large impedance associated with it—at any particular point—back to the main chassis and the board ground.

When considering EMC, it is always good idea for the power integrity engineer to maintain proper practices from the beginning of the design phase, to ensure that noise is mitigated at the very start rather than at the end. Thus, establishing rapport with the EMC engineer at the development phase can help mitigate many EMI issues early on.

6.5 POWER INTEGRITY PDN IN SYSTEM MEASUREMENTS

As seen in the preceding chapters, the *frequency domain* and the *time domain* are the two regimes in which one can measure the quality of a PDN. Both are valuable to an assessment of the efficacy of the power delivery. In Chapter 5 voltage droop was discussed in the time-domain analysis. Measuring voltage droop is the way that PI engineers usually evaluate the efficacy of the PDN, and this is done by inserting dynamic loads into sockets (or in place of where the actual silicon device would be) and then simply measuring the results of a step response on the PDN. Frequency-domain measurements are difficult to do once the system is developed. The PI engineer should make these measurements before a circuit board is populated. However, this data will give only the impedance of the planes, so the value of the measurements is limited, since much of low-frequency behavior is due to the capacitance in the PDN. Another way is to perform a frequency-dependent measurement by using time-domain techniques [29]. This essentially involves having the silicon (or a representative load at the location of the silicon) switch the load current in a certain fashion to create a sort of *impulse response* that allows an *FFT* (fast Fourier transform) to be run on the data to generate an impedance profile. Figure 6.30 shows a simple setup that can be used to run this measurement.

In the figure, at the end of the PDN is a representative dynamic load that indicates the impulse response in the time-domain. Oscilloscope probes near the load pickup signals with bandwidths up to 3 GHz. The scope will capture the voltage waveform during a repetitive set of current load steps based on the changing clock frequencies. The objective is to compute the function $Z(f)$ and compare this with simulated results. As discussed in [30], one starts with the computation of the voltage waveform, which is a convolution of a transfer function and the current function, $i(t)$. Both functions are represented by a Fourier transform as

$$v(t) = \int_{-\infty}^{\infty} h(\tau)\, i(t-\tau)d\tau \;=\; \int_{-\infty}^{\infty} H(\omega)I(\omega)e^{j\omega t}\, d\omega \tag{6.34}$$

To determine the function in $Z(f)$, or the magnitude of this function, the relationship is computed with the measured signal $v(t)$. If an impulse, or delta function, is assumed for the current pulse, the resulting current function becomes

$$i(t) = \delta(t) = \int_{-\infty}^{\infty} I(\omega)e^{j\omega t}\, d\omega \tag{6.35}$$

Since the delta function is a *constant* in the frequency domain, the transfer function is obtained through the transform, as well, as $v(t)$:

$$v(t) = \int_{-\infty}^{\infty} H(\omega)e^{j\omega t}\, d\omega \;=\; h(t) \tag{6.36}$$

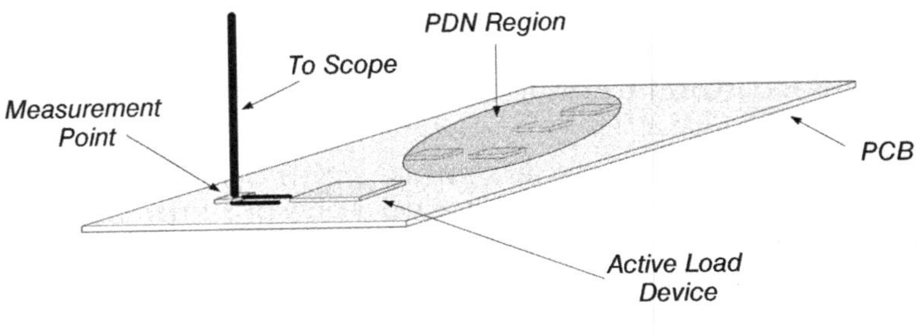

Figure 6.30 Example setup for PDN measurement

which gives the final function

$$H(\omega) = Z(\omega) \cong Z_k = \sum_{n=0}^{N-1} v_n e^{-n(j2\pi k)/N} \tag{6.37}$$

The solution is the desired FFT of the voltage *measured* function. The next thing is to verify that the data matches the simulations. To ensure that the current function is a delta function, the load is set up to run at different clock frequencies to create a reasonable spectrum of currents so represent a wide spectrum from 1 MHz to 1 GHz [31]. There are a number of variations on this approach discussed in the references listed at the end of the chapter. Some call for engineers to construct *load* units that take the place of an actual device. The load unit is a programmable device with active FETs that can switch in *controlled* currents at specific rates and times. The advantage of this method is that the load unit is designed specifically for the purpose of checking the efficacy of the PDN, whereas an actual device is not. If the device is *not* soldered to the board and is socketed, a load unit is inserted into the socket of the actual PCB and used for characterizing the PDN in the actual system. The disadvantage of the load unit is that it is difficult to switch the load at the same frequencies as a high-performance silicon device can switch—thus, its bandwidth is limited. However, the advantage is that both time-domain and frequency-domain data are generated once the measurements are made. Thus, not only the impedance of the PDN is obtained but also the time-domain droop for a given *di*/*dt* at the load.

6.6 SUMMARY

This chapter focused on system considerations that affect power integrity. Though power integrity is typically limited to the PDN in the system, there are system-level issues that must be dealt with to ensure a quality platform design. The loadline was introduced first, illustrating how, in some power delivery systems, a loadline gives more margin at both the high end of the voltage scale for operation and the low end of operation. A good loadline will guarantee reliability and performance throughout the silicon devices range of load. The loadline led to voltage guardbands and the importance of using guardbands on the system to ensure performance. The voltage guardband can accumulate from a number of sources, including the process of the silicon and the design and manufacture of the voltage regulator controller. The guardband contributes to the efficacy of power delivery for high-performance silicon devices. Noise issues were next considered at the system level. Different types of noise were discussed—not only within the PDN—but also at other planes and signals in the system. Noise was observed to be self-generated through the voltage regulator or the load as well as coupled into the PDN from other sources—such as SSN from IO. The additional noise sources could result in the loss of guardbands at the load and thus have to be dealt with at the design phase, from the very beginning. It was then discussed how to address noise using additional high-frequency capacitive filtering and a *tight* ground loop layout around the silicon devices and capacitors. Methods for modeling different noise mechanisms in planes were also discussed, one of which is the segmentation method.

Though EMI is typically not a focus for the PI engineer, it is still an important aspect of the PDN design and power delivery. Thus, a discussion of how EMI can radiate from a system ensued, and how common-mode and/or differential-mode currents can contribute to an EMC problem. Regarding EMC, the PI engineer has a critical responsibility to ensure that the product meets the compliance standards. A few measurement techniques for the PDN were illustrated, once the PDN is designed. Both frequency-domain and time-domain measurements were demonstrated to be useful in determining the quality of the network. An impulse response was used to gather frequency-dependent data while in the system, and this was shown to be a good way to learn where the resonances occur and where droops could dominate in the PDN—particularly if the designers plan on re-doing the PCB layout with the objective of improving the PDN.

REFERENCES

1. Koertzen, H. W. Impact of die sensing on CPU power delivery. *APEC*, 2003.
2. Gupta, V., and Rincon-Mora, G. A. Predicting and designing for the impact of process variations and mismatch on the trim range and yield of bandgap references. *MWSCAS*, 2002.
3. Eirea, G., and Sanders, S. R. High precision load current sensing using on-line calibration of trace resistance. *IEEE PESC*, 2006.
4. Waizman, A., and Chung, C.-Y. Extended adaptive voltage positioning (EAVP). *IEEE EPEP*, 2000.
5. Waizman, A., and Chung, C.-Y. Resonant free power network desing using extended adaptive voltage positioning (EAVP) Methodology. *IEEE Trans. Advanced Packaging*, 2001.
6. Jeong, K., Kahng, A. B., and Samadi, K. Impact on guardband reduction on design outcomes: A quantitative approach. *IEEE Trans. Semiconductor Manufacturing*, 2009.
7. Cui, W., Fan, J., Ren, Y., Shi, H., Drewniak, J. L., DuBroff, R. E. DC power-bus noise isolation with power-plane segmentation. *IEEE Trans. Electromagnetic Compatibility*, 2003.
8. Cui, W., Fan, J., Shi, H., Drewniak, J. L. DC power-bus isolation with power islands. IEEE EMC Symp., 2001.
9. Hu, H., Hubing, T., and Van Doren, T. Estimation of printed circuit board bus noise at resonance using a simple transmission line model. IEEE EMC Symp., 2001.
10. Mao, J., Archambeault, B., Drewniak, J. L., Van Doren, T. P. Estimating DC power bus noise. IEEE EMC Symp., 2002.
11. Hubing, T., Chen, J., Drewniak, J. L., Van Doren, T., Ren, Y., Fan, J., DuBroff, R. E. Power-bus noise reduction using power islands in printed circuit board designs. IEEE EMC Symp., 1999.
12. Liu, Z. H., Li, E. P., See, K. Y., Liu, X. E. Study of power-bus noise isolation using SPICE compatible method. IEEE EMC Symp., 2005.

13. Hueting, J. E., Hijzen, E. A., Heringa, A., Ludikhuize, A. W., Zandt, M. A. A. Gate-drain charge analysis for switching in power trench MOSFET's. *IEEE Transactions on Electron Devices*, 2004.

14. Lopez, T., and Elferich, R. Accurate performance predictions of power MOSFETs in high switching frequency synchrounous buck converters for VRM. *IEEE PESC*, 2008.

15. Matoglu, E., Pham, N., Selli, G., Lai, M., Connor, S., Drewniak, J. L., Archambeault, B., Wang, D., Kuhn, D., Hashemi, R., De Araujo, D. N., Cases, M., Wilkie, B., Herrman, B., Patel, P. Voltage regulator module noise analysis for high-volume server applications. *IEEE EPEP*, 2005.

16. Feng, G., Selli, G., Chand, K., Lai, M., Xue, L., Drewniak, J. L., Archambeault, B. Analysis of noise coupling result from overlapping power areas within power delivery networks. *IEEE EMC Symposium*, 2006.

17. Knighten, J. L., Archambeault, B., Fan, J., Seilli, G., Xue, L., Conner, S., Drewniak, J. L. PDN design strategies III. Planes and materials—Are they important factors in power bus design? *IEEE EMC Society*, 2006.

18. Paul, C. R. Introduction to Electromagnetic Compatibility. Wiley, 1992.

19. Balanis, C. A. Advanced Engineering Electromagnetics. Wiley, 1989.

20. Okoshi, T., Uehara, Y., and Takeuchi, T. G. The segmentation method—An approach to the analysis of microwave planar circuits. *IEEE MTT Trans.*, 1976.

21. Wang, C., Mao, J., Selli, G., Luan, S., Zhang, L., Fan, J., Pommerenke, D. J., DuBroff, R. E., Drewniak, J. L. An efficient approach for power delivery network deisgn with closed-form expressions for parasitic interconnect inductances. *IEEE Trans. Advanced Packaging*, vol. 29, no. 2, 2005.

22. Swaminathan, M., Chung, D., Grivet-Talocia, S., Bharath, K., Laddha, V., Xie, J. Designing and modeling for power integrity. *IEEE Trans. Electromagnetic Compatibility*, vol. 52, no. 2, 2010.

23. Van Valkenburg, M. E. Network Analysis, 3rd ed. *PrenticeHall*, 1974.

24. Paul, C. R. Physical dimensions vs. electrical dimensions, and modeling for power integrity. *IEEE EMC Society Newsletter*, no. 230, 2011.

25. Shi, H., Sha, F., Drewniak, J. L., Van Doren, T. P., Hubing, T. H. An experimental procedure for characterizing interconnects to a DC power bus on a multilayer printed circuit board. *IEEE Trans. Electromagnetic Compatibility*, vol. 39, no. 4, 1997.

26. Hubing, T. H., Drewniak, J. L., Van Doren, T. P., Baudencistal, P., Power bus decoupling on multilayer printed circuit boards. *IEEE Trans. Electromagnetic Compatibility*, vol. 37, no. 2, 1995.

27. Ott, H. W. Electromagnetic Compatibility Engineering. Wiley, 2009.

28. Cui, W., Ye, X., Archambeault, B., White, D., Li, M., Drewniak, J. L. EMI resulting from signal via transitions through the DC power bus. *Electromagnetic Compatibility*: IEEE Int. Symp; 2000.

29. Waizman, A., Livshitz, M., Sotman, M. Integrated power supply frequency domain impedance meter (IFDIM). *IEEE EPEP*, 2004.

30. Since, H. J. H., Jalaluddin, B. Y. A., Yeong, O. A. Y. Study of high speed current excitation reverse engineering methodology using measured voltage and PDN impedance profile from a running microprocessor. *IEEE EPEP*, 2007.

31. Thirugnanam, R., Ha, D. S., Mak, T. M. On channel modeling for impulse-based communications over a microprocessor's power distribution network. IEEE ISPLC Symp., 2007.

PROBLEMS

6.1 A power integrity engineer has determined that the minimum operating voltage for a device is 0.8 V. If V_{max} at zero load is 1.2 V, and I_{max} is 65 amps, determine the loadline. If the loadline error is $+/-$ 10% and the initial voltage guardband is 50 mV, what is the new guardband?

6.2 Compute the TOB using 3σ variance for Table 6.2, assuming that every value has doubled. What is the new TOB?

6.3 Using the transmission line method, determine the noise coupling in Example with the width doubled.

6.4 Estimate the electric field noise from a differential-mode noise source (Section 6.4) if the length of the planes is 10 cm separated by 2 mm at 550 MHz. How large is the noise in dB?

7

SILICON POWER DISTRIBUTION AND ANALYSIS

With the advancement of silicon and integration of multiple chips into single devices and packages, it is now understood that *inside* the silicon is where the *quality* of the power delivery path must be addressed. As the power supply gets closer to the actual load, the power distribution issues that the platform teams have to address, are even more critical when the distribution path is also now silicon. This chapter focuses on silicon power integrity, and how to deliver high-quality power to the load when power comes onto or near the die. Similar to the platform PDN, the problems and analysis of on-die power distribution will be examined. The *on-die* or *on-silicon* link is the electrical path from the bumps, or contacts on the package, all the way to the load. As the issues around the PDN are analyzed, it will be necessary to discuss layout and routing constraints on die as well. The distribution on die and how that distribution routes to the loads are critical to power integrity because these paths determine most of the quality of the power delivery. Additionally, there will be a discussion on basic *on-die* and *on-package* power conversion focused on package regulation and power regulation techniques (with switching power FETs, etc.). There are other internal methods, such as linear regulation and switched capacitor technologies; see also Chapter 2. With the moving of the voltage regulation closer to the load, there are different requirements on the VRs that must be considered—including a new PDN that is developed for this short path and the decoupling strategies on and off die.

Since at micron- and submicron-level routing layers, there are many thousands of signal routes on silicon, the problem becomes more complex with each silicon

Fundamentals of Power Integrity for Computer Platforms and Systems, First Edition. J. Ted DiBene II.

routing layer egressing down to the actual transistors. Thus, much of the discussion in this chapter is on the routing constraints and issues for the PI engineer around this type of power delivery. Next in importance is *on-die* decoupling, which is key to delivering high-quality power to the loads. More than one type of decoupling is used for both the silicon-level PDN and near the load, with the salient types addressed herein. Last included is a discussion of on-silicon power conversion as it relates to the switching converter, and what impacts to future power integrity problems that type of conversion might bring. This discussion is based on the expectation that on-silicon buck-type regulation will eventually become a production-level technology, and hence, the concepts discussed in this chapter would apply.

As devices become more highly integrated, particularly for SoCs (system on a chip) for commercial applications, the need to understand power integrity and power quality on silicon will become a necessity for engineers involved with such designs. The next section introduces these concepts for on-silicon power integrity.

7.1 SILICON AND PACKAGE POWER INTEGRITY

Though power integrity is still a relatively new field of study, power integrity at the *silicon* level is even more novel. In many cases, the platform-level and silicon-level issues are similar, but there are some important differences that should be understood before delving too far into the silicon regime. Table 7.1 lists some of these differences for reference.

The issues change depending on the type of package that is used on the platform, and how the power is distributed into the die, as was discussed at the platform level in Chapter 4. Since the subject has now moved the distribution from the platform onto the *package and* die, there is an expectation that the frequency behavior of the

TABLE 7.1 Some distinctions between silicon and platform power integrity

PI Category	Platform Power Integrity	Silicon/Package Power Integrity
PDN—time domain	First-, second-, and third-order droops evaluated	First- and second-order droops studied
PDN—frequency domain	Frequency range of interest 200 KHz to 300 MHz	10–300 MHz+
Power planes	PCB and package primarily	Package and on-die power distribution layers
Noise analysis	PCB and package	Package and silicon layers
Capacitance	LF, MF, and HF caps onto package	HF package and inter-die capacitance near loads and adjacent to
Emissions	Typically platform and global—far field mostly	Local and near field mostly

power path has also changed. This is by necessity, since the impedance between the source and load is now much smaller and the response to changes quicker. Probably the biggest difference is in the multiple loads that reside on one die and the interaction between them—particularly with a *common-ground* system—and then how that affects the quality of the power at every stage. Figure 7.1 shows an isometric of a die with multiple loads, or SoC, where the multiple power sources are required to power the entire silicon device.[1]

Recall that when considering the power integrity of the complete system, it is required to have a system view of the silicon platform as well. This is particularly important when power is routed from the package and onto the die from different planes and power sources. Figure 7.2 shows how the routes would egress into the silicon for the example in Figure 7.1. The layers of the package, for the functional blocks in the figure, *are cross-hatched* for clarity. Power flows in from the PCB and into each block/load through power planes in the layers of the package. The layers are positioned so that the current can flow with the lowest impedance into the block through package pads, wires, and bumps into the silicon die. The dashed lines represent power flowing in from layers embedded in the package.

For a high-performance device, such as an SoC, the quality of power must be high. The device is made up of blocks that are manufactured on the same process, and thus, the voltage levels (except for possibly the IOs) are similar to each other. The purpose of splitting the planes and using distinct power rails is for power management—for example, in order to turn off and/or modulate the voltage rails independently of each other. Also, because the IO routing typically dominates high-performance packages, the package designer has to consider the routing issues of the signalling along with the power planes to ensure all of the signals escape with the best signal integrity. Additionally, depending on the required power per rail, the thickness and widths of the copper planes from the package will differ, with some being wider or thinner than others.

In the layout of a *real* device, the blocks, or loads, are active at certain times depending on what software, firmware, and modes the device is operating in [3]. When a block is active, that region draws a certain amount of current over time,

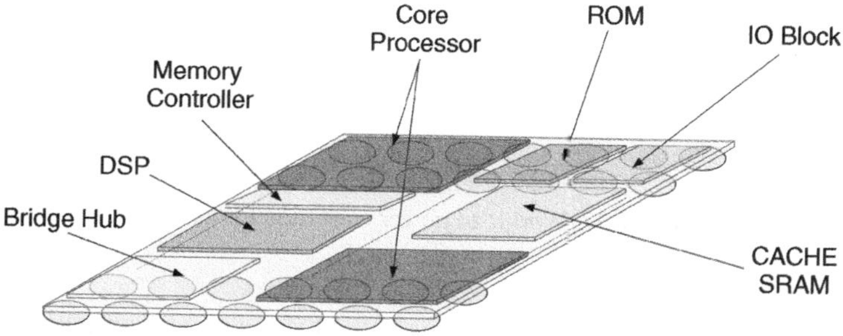

Figure 7.1 Simple isometric of a silicon block with a multiple-load SoC

[1]This picture is for illustration purposes only and does not represent an actual silicon layout or device.

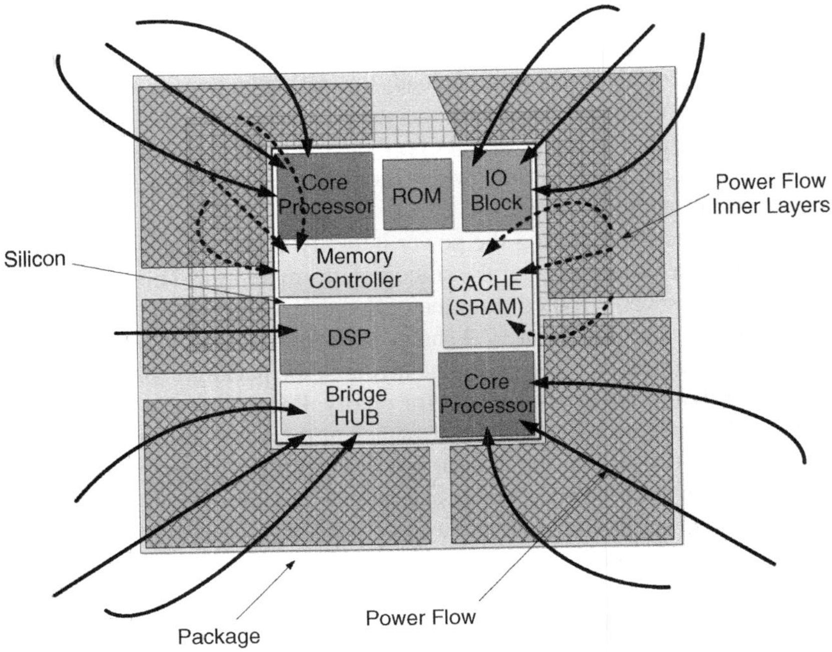

Figure 7.2 Power flow from package to silicon

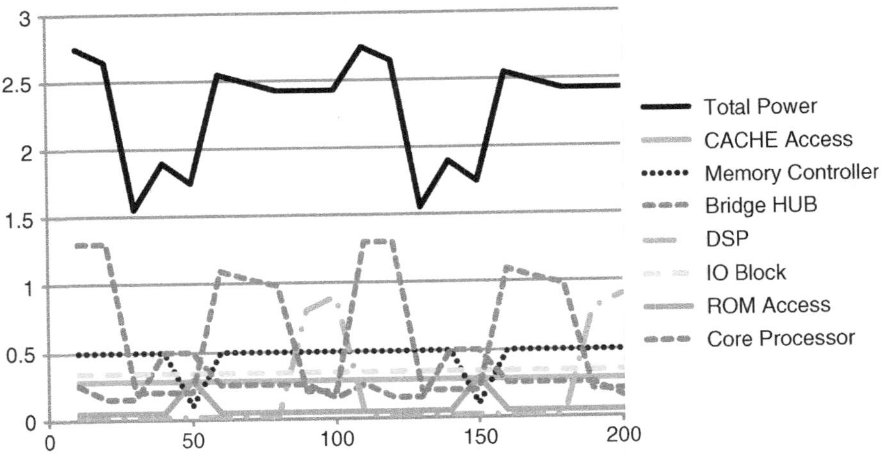

Figure 7.3 Workload power over time for an SoC

which is added to the overall active power of the load over a larger period of time. This is similar to parts of an array of lights *firing* at certain times and lighting up just those regions on a wall. A simple workload chart is shown in Figure 7.3. The workload is plotted as a function of time in milliseconds. The total power at any instance is plotted as the solid black line. Notice that the individual power for each block is plotted as well. The importance of showing the workload for each region is that it helps to clarify where the decoupling needs to be relative to the load—and potentially which load block is most susceptible to noise coupling and reduction of noise margin from the other sources. The other information that is required to understand is how the decoupling need is related to the current rate of change, or di/dt for each load, when it transitions. This is usually associated with the clocks for that block and the measurement of the corresponding current profile.

As was discussed in Chapter 5, this current profile is what shows the efficacy of the PDN for this particular power path. However, in the case of a primarily *silicon* based PDN, how does one know if there is enough decoupling and how does one confirm that they have satisfied the droop requirements for the voltage setpoints in the design? To answer this, it is required to start with the interconnect—and in particular, the distribution of the planes and traces coming from the package and eventually, onto the die. It will be required to take some liberties here though in the descriptions since the power source and placement is critical towards understanding the flow of power into the load. Thus, the objective will be to begin with the power layer structure on silicon and then make our way back towards the power source—be it on-die or on-package. Thus, the next section focuses on power distribution layers on-silicon and then progresses back toward the die interconnection.

7.1.1 Silicon Interconnection for Power Distribution

For many digital (and even mixed analog/digital) devices, there are multiple layers to the silicon construction. How many, depends on the process technology and device, among other items. The intention here is to make this as generic as possible. So an eight-layer *stackup* will be assumed for the moment, and the distribution strategy will be formulated from there.[2] For higher or lower count stackups, the concepts are similar.

In Figure 7.4 the stackup of a device is made up of the substrate (where the transistors reside) and the different metal routing layers, shown from bottom to top in the figure. Note that the device has been inverted relative to the package connections to illustrate the layers and internal connections better.

Also note in the figure that the widths of the traces, and their thicknesses, increase from the lowest layer to the highest layer. This is typical for many manufacturing processes for digital devices [1]. The layers are interconnected through *vias* (not shown in the figure), which connect from the substrate all the way to the topmost layer. Conventionally, the thickest and highest metal layers are used for the power distribution. This is because these layers have the most metal and thus the IR drop across them will be the lowest. Depending on the design, the 3 to 5 topmost layers will be dedicated (primarily) to power distribution, while the lower layers will be dedicated to signals. The trace widths are the smallest in

[2] It is not uncommon for some devices to have 10 or more layers, depending on complexity. Some less complex devices may have 6 or fewer layers.

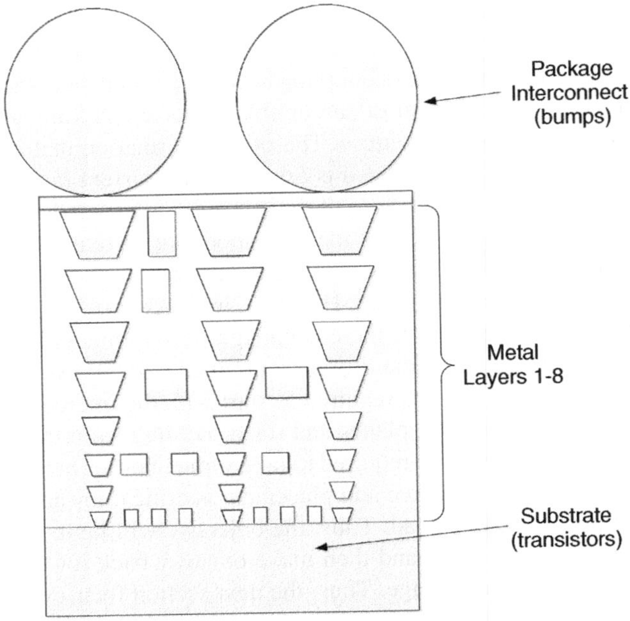

Figure 7.4 Die to bump stackup example

the lower metal layers; this is expected because the signal density is largest here. The signals must make their way up toward the bumps so that they can be routed out onto the package. The primary on-die routing path is toward the periphery of the die where the signals can escape to the topmost layers of the package for routing on the top layers. This is the path for high-speed signals and clocks in particular—mainly for signal integrity and propagation speed reasons.[3] Figure 7.5 gives an isometric showing the arrangement of a few of the layers in the silicon stackup.

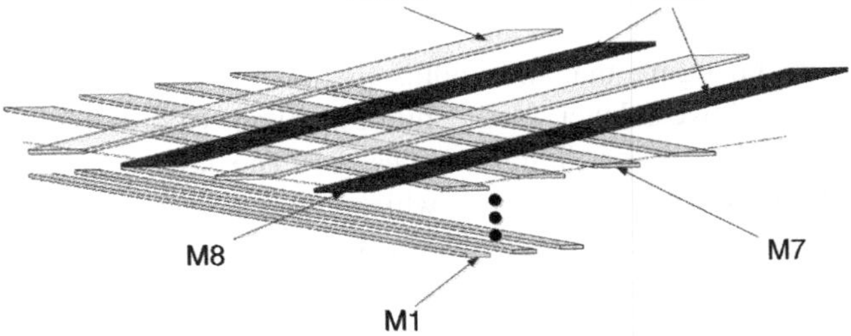

Figure 7.5 Power routes in stacked up silicon layers

[3]The propagation delay on the surface of the package and board is faster due to the microstrip structure being partly in air rather than buried in the layers in a stripline structure. The *effective* dielectric constant for a microstrip is thus lower than a stripline—all else being equal.

The layers alternate in opposite directions for manufacturability, as shown in the figure. The pitch on each trace and width per layer are dependent on which process and what guidelines are set by the manufacturer [2, ch. 7]. Some generic guidelines for the power layers—with respect to power integrity—are as follows: The dielectric between the metal layers, sometimes called *ILD*, or *inter-layer dielectric*, is an insulator. The dielectric material is typically an oxide, such as SiO_2. The thickness of the dielectric layers is in microns or submicrons—which is why, when bounded by two metals, the construction of high-frequency capacitors must include reasonably adequate capacitance values. However, as will be seen for power decoupling, the required capacitance needed is relatively larger than for an *inter-metal* capacitance, and thus, other capacitive structures are often used for this. On a given layer, the power and ground returns are typically routed in pairs to minimize the loop inductance. This is why each layer is directly orthogonal to the routing direction of the layer above it, making it more difficult to maintain a small closed-loop inductive path.

As was discussed in Chapters 4 and 5, the position of the load relative to the power source is an important consideration in creating a quality distribution network. To illustrate the potential issues with routing power into and out of a block region, a processor core here is used as an example in Figure 7.6. As seen in the left side of the Figure, the core is positioned (in this case) in the upper left-hand portion of the processor. The core for a processor can occupy a square millimeter or more of the area, and thus, routing from more than one side is beneficial in keeping the impedance of the path short. The initial thought might be to distribute the power planes equally across the entire core region and then drop the power down into the transistors (substrate) vertically down everywhere. Unfortunately, this is not practical because (potentially) thousands of signals need to be routed into and out of the core region. Moreover, the power/ground plane distribution into the core must be thought out carefully because of a small amount of copper available on the M8 (or in general) upper layers.

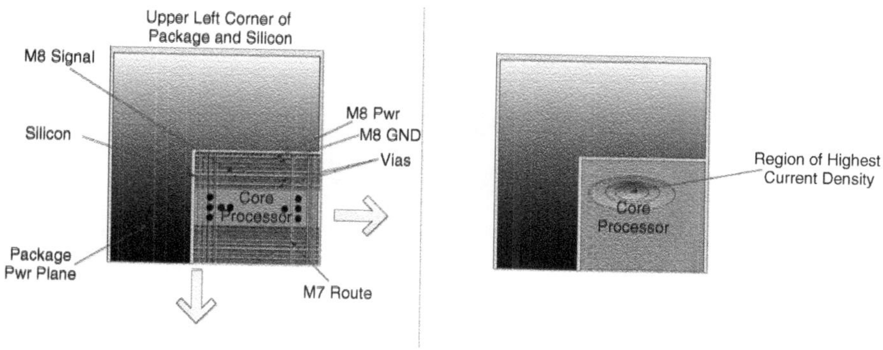

Figure 7.6 Core region silicon layer routing M8 and M7 (left) and dominant current density region on (right)

The minimum *width* of the copper increases as the layers become thicker, and with the addition of certain copper density limits at each layer, it becomes difficult to route large numbers of lines out to the package in any given area near the bump array. Another complexity here is that the current density is not *equally distributed* throughout the core at any instance. Thus, one needs to determine where the largest *average* current density will be and distribute the layers into and out of this region accordingly. On the right side of Figure 7.6, it is evident that the current load density is focused in an area where it is greatest. Typically, as seen in Chapter 5, this is where the processor core *executes* its activity and thus the largest amount of copper, both vertical and horizontally routed, should be in this region—*and* should egress to the package bumps with the lowest resistance possible. Moreover, the layout, or *mask designer* (MD) in this case, should distribute the signals and power into and out of this area according to the current distribution and signal density constraints. Given that this region is replete with signals as well—with the bumps of the package distributed throughout this region—it is almost always a challenge to accommodate all of the connections for an MD. Figure 7.7 illustrates the procedure used to route into this core region and the signal issues (if any) that may require routing out to the package.[4]

A few bump connections are shown for clarity. Note that some of the power and ground traces in M8 are *split* to allow signal connections up to the top layer. However, in the region where the current density is high, the traces stretch all the way across for both M8 and M7. This allows multiple bump connections to the power/ground trace pairs for connections to the package planes and helps facilitate a low impedance path thereto. As the connections make their way to the package,

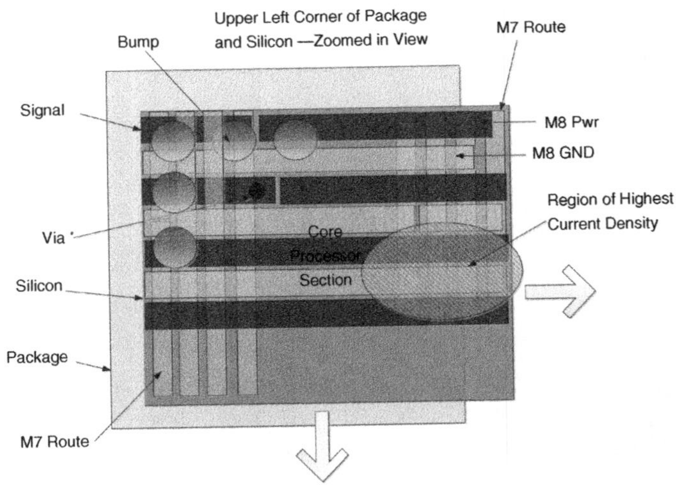

Figure 7.7 Zoomed-in view of power/ground routes in Figure 7.6 showing

[4]For most processing units, the IO is connected to other blocks on-die rather than off-die. The example here is for illustration purposes only.

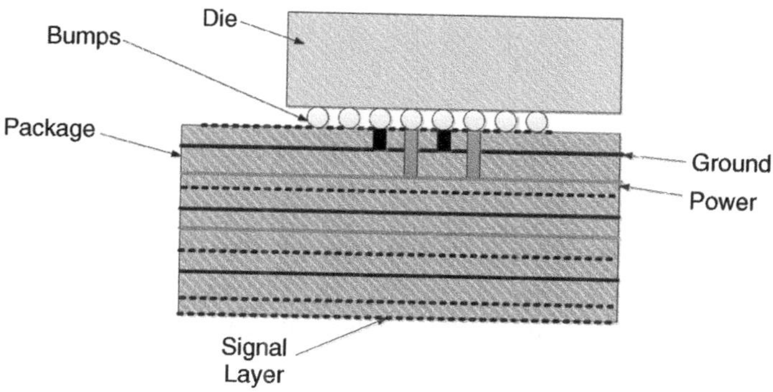

Figure 7.8 Cross section of package and die showing some power and ground connections

it is critical that the power path impedance remain low as well. Figure 7.8 gives a side view of package and die where the layers make their way onto the package through bumps. The path (in this case) represents the connection to an on-package power source where the routing goes into the inner layers of the package.

For the PDN, the power path is represented with discrete elements, and then subsequently each element is analyzed in the path to create the network, as was done in earlier chapters. The important thing to note in the figure is that the objective is to bring down as many connections as possible from the bump profile to enable a reasonable computation for the parasitic elements in the path. The actual resistances for the connections in the path and any current density issues that may occur due to routing improperly are discussed next.

7.1.2 Resistance and Current Density Considerations

From an electrical perspective, it is important that the impedance be as low as possible for the interconnection system between the source and load at the silicon. It would be a good idea to do a first check of the resistance. This path is broken down into three main segments for clarity: the path from the load up through to the bump interconnects, the bump array connections, and finally, the connection to the source on the package. Recall from Chapter 3 that it is possible to determine the resistance of a structure from its geometry. The simple equation for the resistance of a rectangular structure, where the current is uniform, is

$$R = \frac{l\rho}{A} = \frac{l}{A\sigma} = \frac{l}{wt\sigma} \tag{7.1}$$

where l is the length of the trace or plane, A is the cross-sectional area of the segment and σ is the conductivity of the material. This is shown in Figure 7.9.

As temperatures rise, the conductivity of the material changes. Thus, it is important to compute the correct resistance as a function of the conductivity change with

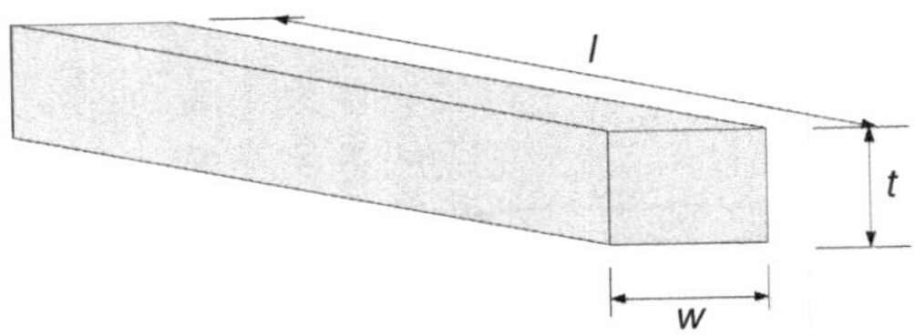

Figure 7.9 Geometry of simple metal strip

respect to temperature. For copper, the conductivity changes in the range of operation of interest here through the following equation:

$$\sigma_T = \frac{\sigma_0}{\left[1 + \alpha\left(T - T_0\right)\right]} \tag{7.2}$$

where σ_0 is the conductivity of copper at temperature T_0, which in this case is room temperature (20°C), and α is 4.3% (0.0043) for copper at the lower temperatures in this case (temperatures given in celsius). Equations (7.1) and (7.2) can be used to compute the resistance for a strip *egressing* from the load back to bump connections (or a group of strips). However, these equations are only applicable if the current is *uniform* throughout the cross section of the strip, that is,

$$J = C = (\text{constant})\ \frac{A}{m^2} \tag{7.3}$$

Thus, the current density must be constant throughout the structure for this expression to be true. However, in many cases this is *not* a valid assumption. The main reason has to do with the position of the components of the power converter relative to the position of the load. As was discussed in Chapter 4, the current density in a plane is not always constant—particularly on the PCB—relative to the load. The position of the VR components—the inductors in this case—can significantly affect the impedance, and thus, must be considered when attempting to extract either the resistance or the inductance. The same holds true for the position of the load relative to the vertical connections. Figure 7.10 illustrates this dilemma.

Depending on the relative position of the source to the load, the current can reverse unequally and flow through vias and bumps as shown in the figure (current increases through the bumps from left to right). The thicker dashed line indicates a higher current density on right verses the left side. This is especially the case where there are significant differences in the resistance and/or impedance of the interconnections as the routes make their way from the load to the source. As shown in Figure 7.10, higher current may flow through the bumps and vias nearest the current source if the package planes, for example, have a much lower impedance than the bumps or via connections in the silicon. This can lead to *electromigration* problems, as will be discussed in detail shortly. Recall from Chapter 4 that the position of the inductors has an effect on the resistance and impedance of the planes relative to the load. What if the VR is now placed on the

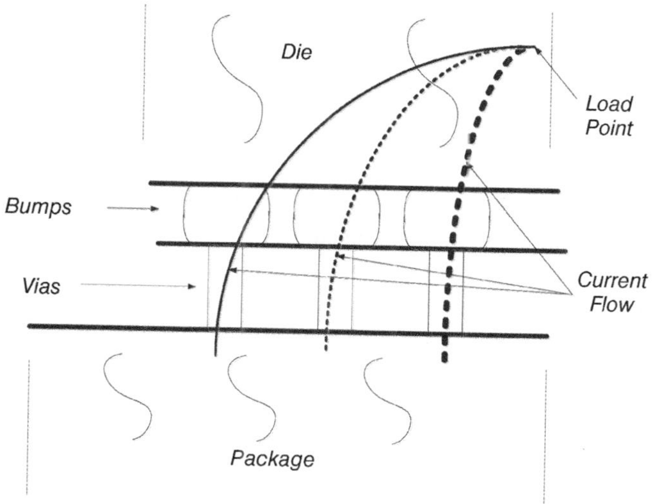

Figure 7.10 Current density changes from load point in silicon

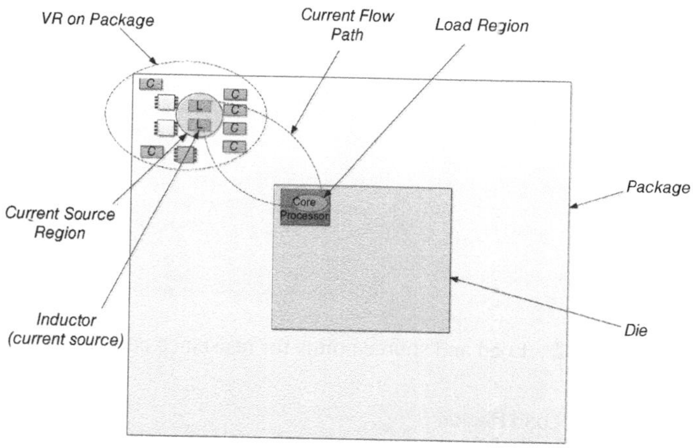

Figure 7.11 VR on the package current path

package? Does it have a similar effect? Figure 7.11 shows a possible VR structure and how it can affect the parasitics of the PDN.

The components in the upper left-hand corner are positioned as one would expect the flow of current to occur in a basic buck design (in this case powering the core unit), with the input capacitors, the power devices and controller, and then the LC filters placed accordingly from left to right. The actual current flows through a contiguous plane on the package toward the silicon load from the actual contacts of the output of the inductors. However, the two outputs of the inductor can also be represented as one-*point source* as was done in Chapter 4 for the platform VR.

This allows the use of the formula for current flow on a plane from two vias (see Chapter 3) to estimate the resistance through the plane to the bottom of the bump connections to the load source on the die. This path is broken down in the following example and each segment analyzed to determine the best estimate for the resistance of the link.

Example 7.1 Refer to the diagrams in Figures 7.12 and 7.13 for the current path from the VR to the bumps, and the dimensions in Table 7.2. Determine the approximate resistance using equation (3.121) in Chapter 3. Assume that the current from the load region is approximately uniform and the current outside of this region is negligible. Also assume that the bumps are distributed so that the current distribution appears to be uniform throughout the specified region for the bump array. The

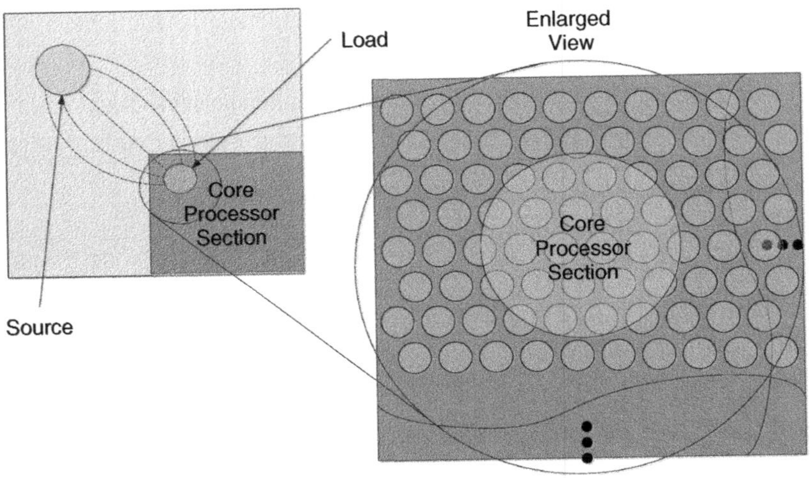

Figure 7.12 Load and source points for resistance computation

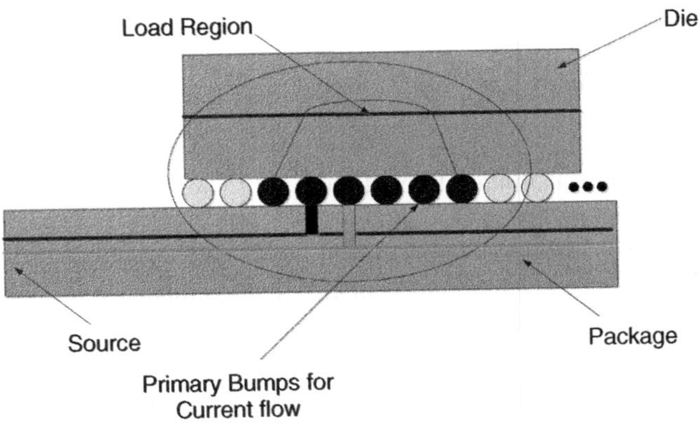

Figure 7.13 Cross section of die/package region in Figure 7.12

TABLE 7.2 Values for computation of resistance from load to source

Element	Dimension
Diameter of source region (inductors)	2 mm
Diameter of load region (processing unit)	0.5 mm
Center to center spacing from source to load	6 mm
Thickness of power and ground planes	0.7 mil
Number of pwr bumps	10
Number of ground bumps	10

temperature of the planes is approximately uniform and resides at 80°C. Ignore the resistance of the vias in the package for this calculation.

If the resistance of the bump array is 4.4 mΩ, and the resistance to the load through the vias is 100 μΩ, is the approximation reasonable? That is, can one assume that the current will flow uniformly through the bumps in the array?

SOLUTION: Recall the equation for the current between two vias in a plane, that is, equation 3.120 in Chapter 3. This result is only valid if the radii are equivalent. The more general result needs to be determined. Assume that the spacing is much larger than the radius of the largest circle. First, designate R as the larger of the two radii and r as the smaller one. Because the current is uniform, the current and *current density* are constant; thus, using equation (3.120) yields

$$I = \oiint J_{\phi} \cdot dS = \int_{r}^{s-R} \int_{0}^{t} \frac{V\sigma}{2\pi r}\, drdt = \frac{V\sigma t}{2\pi} \ln\left(\frac{s-R}{r}\right) \tag{7.4}$$

And for the resistance,

$$R = \frac{V}{1} = \frac{2\pi}{\sigma t \ln\left(\frac{s-R}{r}\right)} \tag{7.5}$$

The limitations on these two equations are that the separation s must be approximately 5× the radii or larger to be valid. A simple check verifies that this appears to be so.

Now, estimate the actual resistance of the plane given the thickness t, using equation (7.5) above. Because here the total path is being considered, the return portion of the path must be multipled by 2. It is important to use the proper copper *conductivity* due to temperature effects, which is

$$\sigma_T = \frac{\sigma_0}{\left[1 + \alpha\left(T - T_0\right)\right]} = \frac{5.7e^7}{[1 + 0.0043\,(80 - 20)]} \cong 4.53e^7 \tag{7.6}$$

$$R_{path} = 2 \times \left[\frac{2\pi}{\sigma t} \ln\left(\frac{r}{s-R}\right)\right] = 2\frac{2\pi}{(4.53e^7)\,t\, \ln\left(\frac{6e^{-3}-1e^{-3}}{2.5e^{-4}}\right)} \cong 5.3\ \text{m}\Omega \tag{7.7}$$

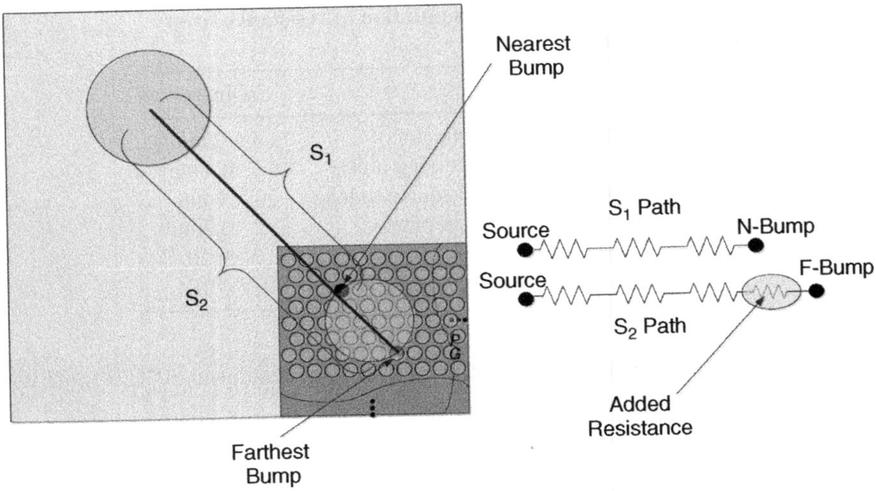

Figure 7.14 Path differences for bumps in array on silicon

This value is a reasonably low given that the planes are 0.5 oz copper.[5] Now, the objective is to determine if the current will flow through the bump array somewhat uniformly. The only way this can be shown is if the resistance of the planes is much greater than the bump resistance. Since the bump resistance is on the same order of magnitude as the plane resistance, the assumption is not valid. The other issue is that the resistance to the actual transistor loads is very low, which means that one can essentially ignore this resistance and assume that the load is co-located at the bumps. These two pieces of data lead to the conclusion that some of the bumps will draw more current than others. One way to visualize this effect is to compare the path from a bump on one side of the array with the path on the other side. Figure 7.14 shows this graphically. If the source is to the left and the resistance per bump is relatively high (in this case, one can assume the resistance to be approximately 10× higher due to the number of power/ground bumps), it is clear that the current path from the far right side will be higher than the one on the far left.

Thus, one might expect, the resistance would increase as a function of the additional plane resistance in getting from a bump on right side to the left as shown in the resistor diagram on the right-hand side of the figure. The good news is that the difference in resistance should be relatively low, given that the overall difference in distances to the source from each bump is small relative to the overall distance:

$$\frac{S_1}{S_2} = \frac{6}{6.25} = 0.96 \tag{7.8}$$

[5]Many package and board planes are specified in ounces (oz), rather than inches or millimeters: 1 oz copper is typically 1.4 mils or roughly 35 microns thick.

Since the plane is uniform and the distance is large relative to the radii, the resistance will be almost proportional to the distances for each bump. Thus, though there is a difference, the amount of current per bump will not increase significantly due to the location and the resistances in the plane. Nonetheless, it is a good idea to check the current densities in interconnection systems, since there is the potential for long-term failures due to changes in resistance over time. Note that only one direction in the power path was checked—the voltage path. Often, the power and ground paths are unequal in resistance and need to be checked separately.

The preceding example illustrates the importance of checking current densities in an interconnect. Though the ratio of current from two extreme bumps was close to one, what if the ratio was much less? What if much more current flowed through one or more bumps or interconnects than the others? Though this is not the best use of the interconnects—for example, the power per interconnect is not shared equally—it can lead to serious reliability problems long term [5]. *Electromigration* is a phenomenon that can happen to metal interconnects if the current densities become much higher than the reliability ratings of the connections allow. Technically, electromigration is the transport of a material over time in a metal caused by the movement of ions in the conductor. The material may move as a result, creating a boundary that can cause a fracture at that location and ultimately electrical failure. The silicon industry uses many methods to predict this problem, since in many advanced silicon devices where current densities are high, the device interconnects are particularly susceptible to this problem. *Black's equation* was developed in the 1960s to help predict the reliability of an interconnect and its eventual failure point [4]. The mean time to failure, or *MTTF* of a wire is computed as,

$$MTTF = \frac{A}{J^n} e^{E_a/kT} \tag{7.9}$$

where A is a constant, determined from the cross-section of the area through which the current flows, J is the current density, n is the scaling factor (typically 2), E_a is the activation energy (0.55 eV for solder and ~0.7eV for copper—both *self-diffusion* constants), k is Boltzmann's constant, and T is temperature (Kelvin). Equation (7.9) can give a reasonable *estimation* for the MTTF of an interconnect. However, in today's systems, electromechanical modeling with 3D and 2D tools are used to model potential interconnect EM problems in actual circuits due to the complexity of advanced silicon and interconnect systems. For the power integrity engineer, this allows for a gross check on the electrical interconnect out to the source to ensure that the design of the PDN is adequate for both PI effects and reliability. This is more of a concern when there are competing elements for limited bump connections to packages, for instance, and the location of the power source—and thus where the resistance thereto—becomes problematic.

7.1.3 PDN Considerations for On-package VR Systems

For the PI engineer involved with on-package voltage regulation, the main goal is to construct an accurate PDN for the power delivery of the die. General system issues must be addressed while developing the PDN to prevent potential failure modes. Once this is understood, working in concert with the package layout engineer, the MD for the silicon, and the VR designer, the PI engineer should be able to formulate the construction of the PDN. Because there are tight routing constraints for devices, such as SoCs, there are limitations in the layers of the package that are available. Moreover, as seen in Figure 7.11, if the devices are placed in a region close to the load, there will be little room on the surface—near the devices and silicon—to route high-speed IO, which limits the package designer's options for surface routed traces. The other problem is with not having *contiguous planes* from the output of the inductors to bumps on the die. Though most packages for such silicon are built-up constructions—meaning that via interconnects are mostly embedded in the stackup—in order to pass signals from the board to the package, it is nearly always required to pass signals vertically through power and ground planes to allow them to be routed from the silicon outward. This is always the case when the die are much smaller than the package and when the IO dominates the package-to-board connections. Thus, when computing the resistance of the planes, a *perforation factor* (PF) is applied to a region of the planes where significant *pass-through* vias occur. The perforation factor is simply an estimate of the actual percentage of copper that is removed in the planes relative to that region—which is then used as part of the final computation for the resistance. Figure 7.15 shows two such regions, one being a contiguous plane and the other with a large number of via interconnects. It is straightforward to estimate this effect: the engineer computes the size of the via *annular* ring (keep out for the pass-through via) and then estimates how many via per area is in that region. An error factor must be applied to ensure that the resistance computation is higher than the actual.

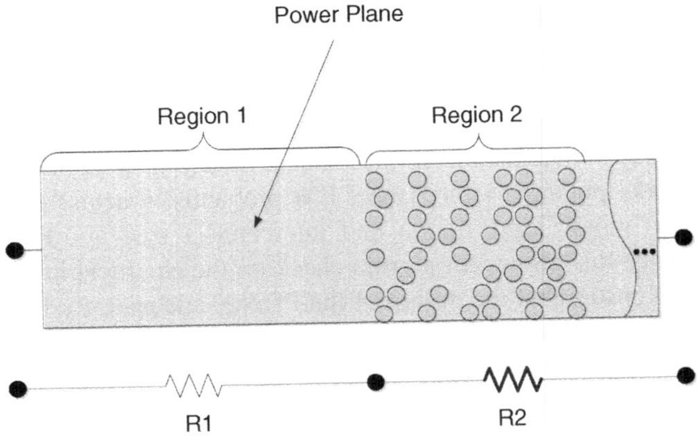

Figure 7.15 Via distribution on plane and resistance difference

As shown in the figure, the via distribution is staggered or compacted close to one another. The issue with using a crude perforation factor is that it does not account for *current crowding* effects in the metal between the tightly crowded vias. The PI enginer should check to see if there are any electromigration issues that could emerge. Additionally, a tightly compacted region can result in higher resistive paths than would otherwise exist if the vias are spaced farther apart. This can occur in packages where pin fields exist of large arrays of vias exist and are consistent through the plane of interest. In such cases, engineers often use *striping* to estimate the planes rather than a PF (perforiation factor), as shown in Figure 7.16. No matter which method is used, the objective is to obtain a reasonable estimate of the resistance that is essentially the same. Moreover, if there is any concern that a reliability problem may exist, the PI engineer should always check with the proper manufacturing representative.

Recall from Chapter 4 that *lumped* elements can be used to estimate the PDN in a first-pass profile. This is not intended to replace numerical modeling, but to supplement it, so that it is possible to reduce the number of iterations in the numerical modeling and get the actual results from either a *first principles analysis* or an estimate that allows the engineer to model a structure to obtain a result that can be used as a comparsion. The first question to ask is whether the PDN is from an on-die VR or a truncated version of the network for the platform? This is a reasonable question to ask, since it is tempting to take a shortcut by simply removing the components from the original estimate for the entire platform, and then use those starting at the end of the VR on the package. In some cases, for a crude estimate, this might be valid. However, the issue is that on-die voltage regulation and power distribution are somewhat different from the platform structures. This difference is mainly in the type of capacitors used for filtering the on-die VR. There are manufacturing limitations as to what type of devices can reside on high-performance packages—most have to do with thermal constraints. For example, the ambient package temperatures are often much higher than the platform board temperatures, and since a heat sink may reside on top of the silicon device, there may also be height constraints on the devices. This means the parts are larger in the X–Y direction, which can

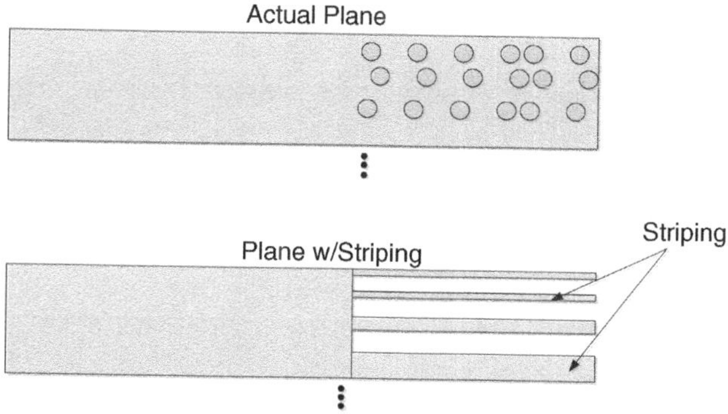

Figure 7.16 Example of striping used to represent planes for resistance computation

increase their placement footprint, and in many cases, their temperatures due to conduction from the package. Thus, the type of components used for such a design is different from those for the board, which will change the characteristics of the PDN somewhat. Additionally, the routing of the devices can change the characteristics of the planes beneath. To understand these effects and those on the PDN for an on-die structure, it is best to start with a simple test case and derive the PDN for it using Figure 7.11, and then build up a PDN, as done in Chapter 4 for the platform version. Figure 7.17 shows how a distribution can be broken down with an on-die structure that is similar to that in Figure 7.11. Note the labeling of each section. There are a good number of elements added here as compared to the platform version(s) illustrated in Chapter 4. These elements are defined in Table 7.3. The main additions are the capacitors that affect the results due to their high-frequency nature, the vertical and lateral route parasitics in the silicon, and the bump impedance.

The cross section in Figure 7.18 shows these elements in more detail. The labels correspond to the schematic in the Figure 7.17. Because of the placement of the load, both the vertical and lateral sections are considered in the silicon routing. The inductive portion, which is often quite low, is ignored. The decoupling on-die has a very low parasitic inductance (as will be shown later in the chapter) and thus is absent in the load capacitance model in Figure 7.17.

There are a few items to note in the figure. As was discussed in Chapter 4, the return path is a mirror image in the model. This indicates that the parasitics of the resistive and inductive elements are approximately the same. In most cases, *this may not be a valid assumption*. The PI engineer needs to get more detail up front if a more accurate estimate is required. However, the details could be left for modeling with 3D tools, when more accuracy in the modeling effort is required later.

The construction of the model is simple enough that the lumped elements will yield a good approximation to the actual. Once the PDN is constructed, a simple Mathcad file could be run to determine the overall impedance profile, or a SPICE model to determine the effects due to droop in the time-domain as was done in Chapter 5. In the next example, a PDN file is constructed to examine the frequency-

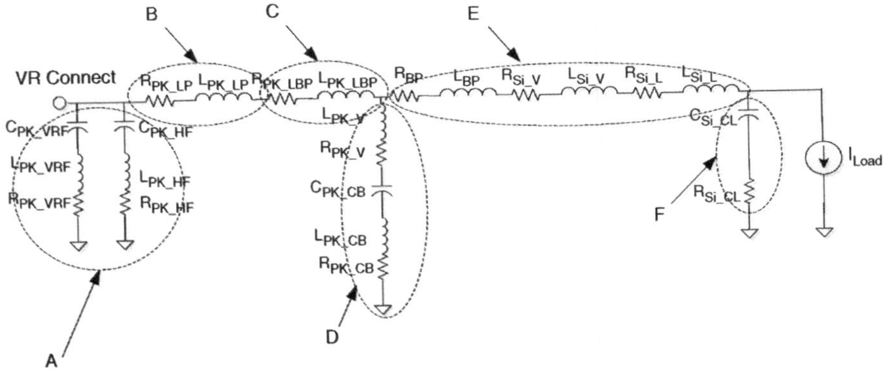

Figure 7.17 Schematic for an on-die VR PDN to load

TABLE 7.3 Description of PDN elements in Figure 7.17

PDN Element	Definition
C_{PK_VRF}	Package-level VR filter capacitor capacitance
L_{PK_VRF}	Package-level VR filter capacitor inductance
R_{PK_VRF}	Package-level VR filter capacitor resistance
C_{PK_HF}	Package-level high-frequency capacitor capacitance—outside of bump region
L_{PK_HF}	Package-level high-frequency capacitor inductance—outside of bump region
R_{PK_HF}	Package-level high-frequency capacitor resistance—outside of bump region
L_{PK_LP}	Package-level plane inductance—outside of bump region
R_{PK_LP}	Package-level plane resistance—outside of bump region
L_{PK_LBP}	Package-level plane inductance—inside of bump region
R_{PK_LBP}	Package-level plane resistance—inside of bump region
L_{PK_V}	Package-level plane inductance—vertical route from backside of package (if any)
R_{PK_V}	Package-level plane resistance—vertical route from backside of package (if any)
C_{PK_CB}	Package-level high-frequency capacitor capacitance—backside of package (if applicable)
L_{PK_CB}	Package-level high-frequency capacitor inductance—backside of package (if applicable)
R_{PK_CB}	Package-level high-frequency capacitor resistance—backside of package (if applicable)
L_{BP}	Bump inductance
R_{BP}	Bump resistance
L_{Si_V}	Silicon-level vertical inductance
R_{Si_V}	Silicon-level vertical resistance
L_{Si_L}	Silicon-level lateral inductance
R_{Si_L}	Silicon-level lateral resistance
C_{Si_CL}	Silicon-level load capacitance
R_{Si_CL}	Silicon-level load capacitance resistance

domain effects and locate the resonances of the structure. The first step is to put together a table of values based on some *first principles* extractions. For the bump array, it is assumed that there is some mutual coupling between the bumps to more accurately determine the inductance. However, as will be seen, this inductance will be very small, and thus, some components will be neglected, or added in series with other elements, to simplify the analysis.

Example 7.2 Using Figures 7.17 and 7.18, and Table 7.4 determine the PDN for the network of Figure 7.17 and plot the magnitude of the impedance. Compute the

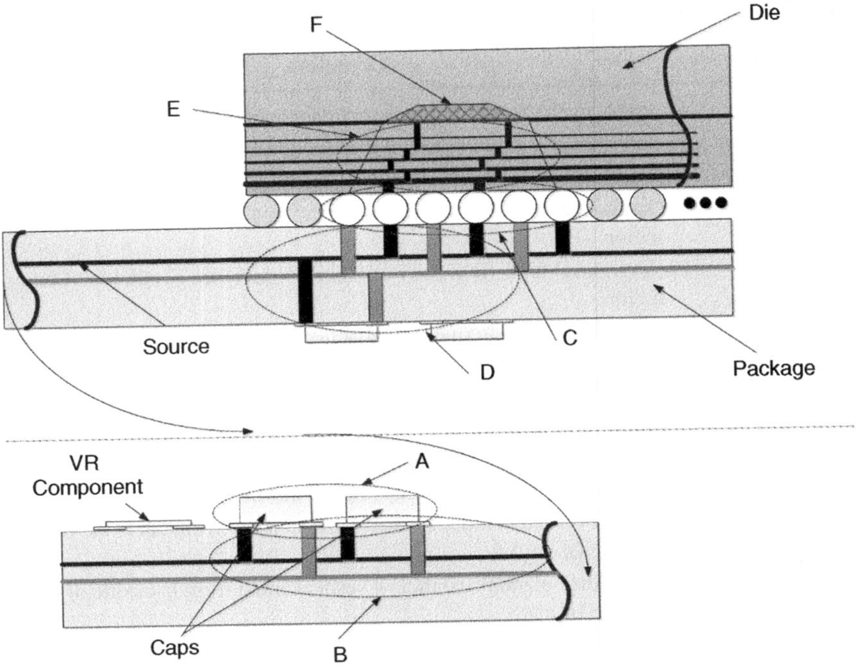

Figure 7.18 Cross-section of PDN regions in Figure 7.17

bump inductance and resistance from the values below. Can one lump this into the other section or ignore it? In other words, are the values large enough that someone should be concerned about them?

Bump computation data:

- Bumps on 180 micron pitch
- Size of bump: 110 micron diameter
- Material: solder ($5.7e^6$ S/m)
- Number of bumps for power and ground (each): 20

SOLUTION: It is best to start by estimating the bump resistance and bump inductance. One can use the fact that the bumps are bounded on both sides by planes, so fringing effects are ignored. Normally, in such an array it might be necessary to compute the *mutual* inductance as well, to see that there is an effective overall inductance on the bump array that is smaller than the original estimate. However, since this effect should be within the error range of the estimate, it is ignored. If the bumps are essentially cylinders, the resistance estimate is straightforward:

$$R = \frac{l\rho}{A} = \frac{d}{\pi r^2 \sigma s} = \frac{2}{\pi r \sigma s} = \frac{2}{\pi \left(55e^{-6}\right)\left(5.7e^6\right)} \cong 2\ \mathrm{m}\Omega \qquad (7.10)$$

TABLE 7.4 PDN values for schematic in Figure 7.17

PDN Element	Metric
C_{PK_VRF}	30 μF
L_{PK_VRF}	300 pH
R_{PK_VRF}	400 μΩ
C_{PK_HF}	10 μF
L_{PK_HF}	44 pH
R_{PK_HF}	100 μΩ
L_{PK_LP}	360 pH
R_{PK_LP}	5.3 mΩ
L_{PK_LBP}	70 pH
R_{PK_LBP}	2 mΩ
L_{PK_V}	0
R_{PK_V}	0
C_{PK_CB}	0
L_{PK_CB}	0
R_{PK_CB}	0
L_{BP}	Bump inductance
R_{BP}	Bump resistance
L_{Si_V}	500 fH
R_{Si_V}	600 μΩ
L_{Si_L}	730 fH
R_{Si_L}	1 mΩ
C_{Si_CL}	20 nF
R_{Si_CL}	400 μΩ

Since there are 20 bumps, the resistance will be divided by 20 and then doubled for each direction, or 200 μΩ for the bump array.

The inductance can be computed using equation (3.115) in Chapter 3,

$$L = \frac{\mu_0 l}{\pi} \ln\left(\frac{(s-r)}{r}\right) = \frac{(4\pi e^{-7})\ (110e^{-6})}{\pi} \ln\left(\frac{180-55}{55}\right) \cong 36 \text{ pH} \quad (7.11)$$

Since there are 20 pairs, the effective inductance is around 1.8 pH. Note that both the resistance and inductance are very small numbers, but they are of similar magnitude to the silicon-level parasitics for the interconnects to the load. This means they should be added together with the other parasitics rather than ignored.

The next step is to put together a Mathcad spreadsheet to determine the impedance profile for the PDN. Figure 7.19 shows the plot for the PDN. Note that there is a large resonance at around 47 MHz. The data are in log scale for both the x and y axes. This indicates that if one were *designing* an actual PDN for this system, it would be desirable to reduce or *shift* out this resonance by adding filtering. One

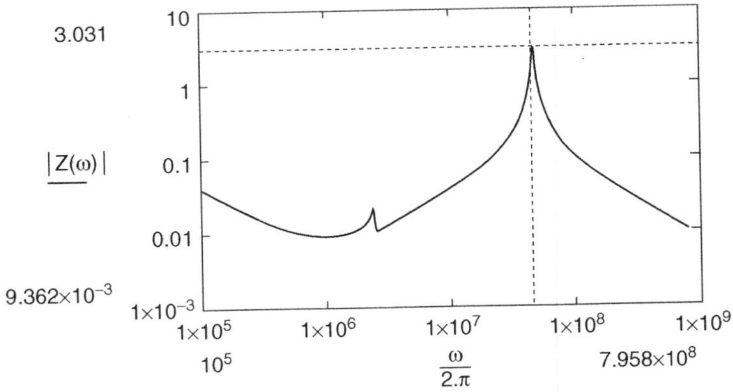

Figure 7.19 PDN Impedance plot for Example 7.2

possible cause of this high peak is that there is no *backside* capacitance added, or rather, there is not enough high-frequency capacitance added with low inductance to the package. The frequency of the resonance may become a source of noise at some point, so it would be a good idea to squelch this.

The other interesting characteristic to note about the plot is the low-frequency impedance. The impedance starts to increase from around 1 MHz as the frequency goes down. This indicates that the loop response should be faster than the point where the impedance starts to rise at a substantial rate. There is a very small resonance in the 3 to 4 MHz range as well. Because the resonance is small and does not align to any of the frequencies of the signals (in most cases), it should not be a concern. As will be shown in the next section, the frequency of operation for the converters is higher than that for the platform. This has to do with the size of the components. The components need to be small enough to fit into a constrained space. Thus, one could increase the switching frequency to accommodate this.

The preceding example showed that the PDNs for on-package applications are different from those for on-platform. In the next section power delivery for both on-die and on-silicon applications are examined.

7.2 SILICON AND PACKAGE POWER DELIVERY

Some basic power integrity issues for silicon- and package-related delivery systems have been discussed up to this point. Some salient design issues and parameters for on-package and on-silicon power delivery are introduced in this section. Relative to basic platform power delivery methods and conversion technologies, this is a new and exciting field of technology that has been gaining interest over the past decade. As was discussed briefly in Chapter 1, the voltages that are being delivered today, along with the high current densities, are driving voltage regulation closer to the silicon level, and in a number of cases, engineers are designing VR's right onto the silicon itself. For the most part, the on-silicon power delivery schemes are relegated to linear regulation and nonswitching designs. Just about every type of voltage regulation method has its place—however, some methods are more suited to certain applications than others are. This section discusses these methods in brief. To do the subject justice would require a complete text, which is beyond the scope here. For the power integrity engineer, understanding the direction of power delivery moving forward is fundamental. The PDN for near silicon or *on-silicon* power regulation schemes brings with it a whole host of new problems. Some of the issues and basic technologies around on package regulation will be discussed in this section and then a few on-silicon power delivery methods within this context. A, brief revisit to the effects on power integrity one can expect from adoption of these methods will be presented last.

7.2.1 On-package Power Delivery

With on-package voltage regulation, as with all new and developing technologies adopted into real systems, come a multitude of trade-offs. From the power integrity perspective, it was shown in the previous sections that the PDN for silicon may be considerably different than the PDN developed for a platform. Primarily, it has to do with the placement of ever smaller components into a very constrained area. There are also some practical environmental issues that must be dealt with when designing such structures. First, the package *plane* temperatures (and thus ambient temperatures typically) for the components are 20°C or hotter than on the platform. This is due in part to the higher junction temperatures of the load device and the proximity of the VR components to that silicon. It is common for such devices on the package to experience temperatures of 90°C or more. Moreover, some special components, especially the FET's and inductors, require elevated internal operating temperatures as compared with their platform counterparts. Second, mechanically, all these components are smaller. This limits the ability of the designer to allow heat to dissipate easily through the planes or through any other mechanism. Third, the voltages generated and brought into such voltage regulators are typically much lower than those of the platform. This is because small builtup layers and materials of the packages are often subject to voltage breakdown problems at the higher voltages over the life of the product. Such constraints can make placing an on-package voltage regulator challenging.

Still, the reasons and advantages for many applications are compelling. Size, reduction of cost due to less platform components, real estate benefits to the platform, and potential performance benefits are just a few of the reasons that have driven designers down this path. From a voltage regulation topology perspective, designers have multiple choices. If the currents are relatively high, the *buck regulator* is the best choice. This was reviewed in some detail in Chapter 2. The main difference here between the platform designs and on-package designs is integration. Figure 7.20 illustrates this.

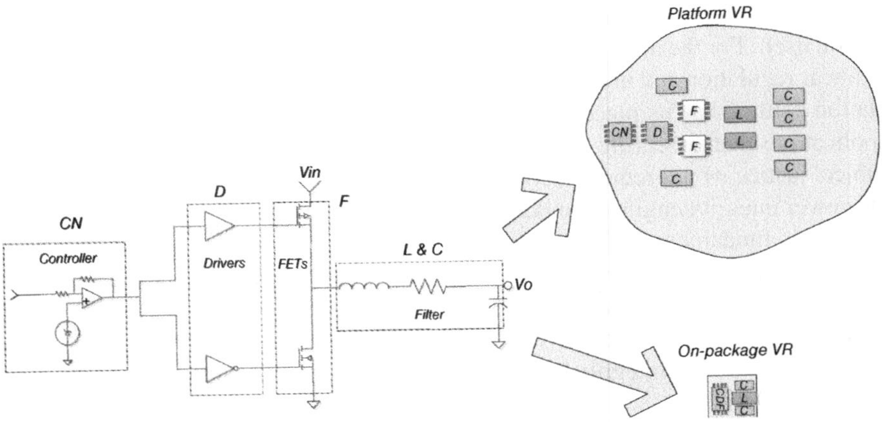

Figure 7.20 Platform and on-package voltage regulator compared

7.2.2 Package and On-Silicon Power Delivery Trade-offs

In the platform case, the FETs, drivers, controller, and filter components are discrete and placed independent of each other. However, in an on-package design, most, if not all, the components is integrated into just a few components. In the case of the FETs, drivers, and controller, these blocks are integrated into one monolithic device. This has the clear benefit of reducing the parasitics between these functions as well. The addition of the substrate allows for ease in assembly onto the package. There are some practical limitations that must be considered, and which can affect the PI engineer. First, the efficiency of the solution is typically lower than that of the platform version. This is because the switching frequencies required to keep the components small can be 5× to 10× greater (or more) than that of the platform. The switching losses in the transistors, as was shown in Chapter 2, are directly related to the frequency and the capacitance (primarily the gate) of the primary switching devices. Thus, one would expect the overall power losses to be somewhat higher than in an equivalent platform design. Moreover, the inductor materials, for discrete inductors, invariably have higher losses at the higher frequencies (ferrites and alloys combined). This further exacerbates the power loss problem and can affect the temperature in the planes of the package, adding higher losses in the PDN and thus higher resistance. If the voltages are low enough, the developers of the SoC, or silicon load device, may choose to incorporate some of the VR components into the SoC. Then again, this may not be the best economic or practical choice due to myriad constraints with other digital blocks in the load die. The additional heat from the power elements can even further exacerbate already existing thermal issues.

The topology of the buck converter design is virtually identical to those of the platform variety.[6] Thus, it will not be necessary to review the circuits again

[6]The purpose here is not to design an on-package or on-silicon voltage regulator but to illustrate the use case and power integrity concerns.

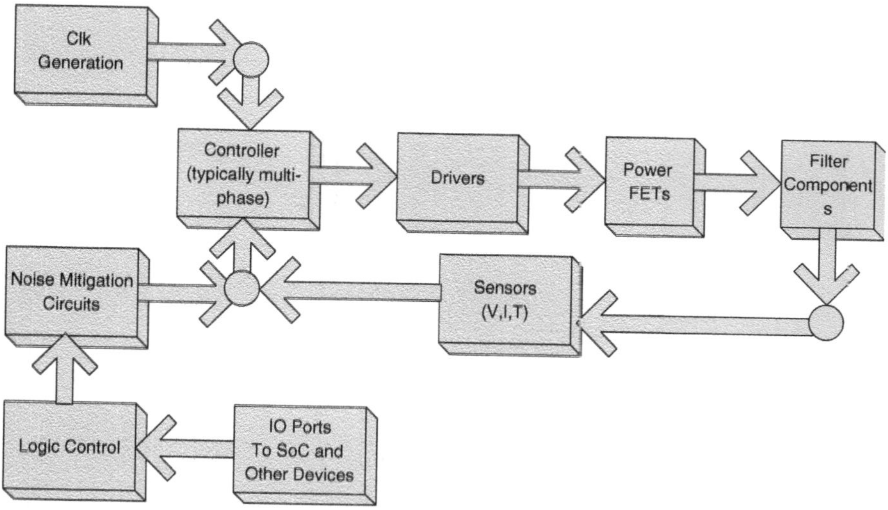

Figure 7.21 Typical blocks for power system integration of-package/on-silicon designs

here—only the differences. Nevertheless, it is practical to illustrate some of the salient attributes in block form. Figure 7.21 shows a design broken down into its main block componentry. The differences between this design and one, say, for a platform of similar power, is that the circuitry required to switch at a faster rate, in order to keep the component sizes small and the amount of integration [6], also requires the circuitry around it to support these speeds. When integrating a number of analog and digital blocks into a single monolithic device, it is often important for designers to also limit the noise coupling between key sections in order to ensure proper operation. Thus, often noise mitigation circuits are added to help reduce any additional noise. Moreover, since relatively high currents are switching in such an integrated voltage regulation device, any device that shares the same planes, IO, and interconnects is susceptible to this noise. Thus, mitigation techniques are crucial when designing highly integrated analog systems.

For lower power applications, there are some other choices for voltage conversion. The *switched* capacitor and the linear regulator are used extensively for many low-power delivery systems—often with integration into the load die. As was shown in Chapter 2, these converters have the advantages of not requiring an inductor as a filter device, and the power delivery is completed with very few discrete components external to the silicon. Figure 7.22 shows where such converters are incorporated into the load die as an example.

The *switched capacitor* design was mentioned in Chapter 2 (charge pump) along with the *linear* regulator or low-drop out linear regulator. Both converters are easily integrated into the existing silicon and often are done so for power applications that require small current loads—for example, PLLs (phase-locked loops), bias power sources, or small logic regions such as IO, which require different voltage

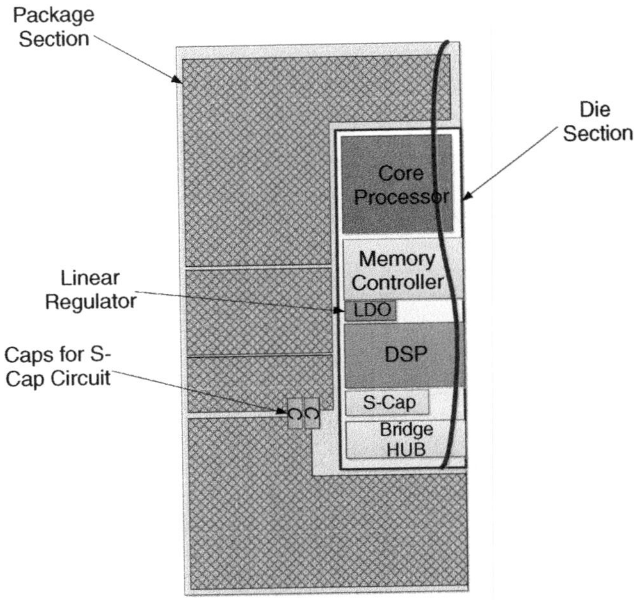

Figure 7.22 Section of die illustrating LDO and switched capacitor

levels than, say, the processing units or memory. Many of these designs can handle currents in the range of a few 100 mA without forcing too many constraints on the digital circuits and layout around them. The benefit, of course is that this removes the need for external devices. When a complex SoC is being put together, there is almost always a need for multiple small rails to support internal functions. Both the switch-capacitor design and the linear regulator offer benefits of size and flexibility for such applications. Where these converters fall short is when the application requires (in general) one or more of the following attributes:

- Currents larger than 500 mA
- Relatively large output voltages requiring a different transistor process
- Wide static or dynamic voltage regulation
- Large dynamic current changes

Linear regulators also require voltage headroom, or a voltage *delta*, that is large enough to allow the linear to regulate and turn on. This is because a linear regulator has a limited operating range if the source and output voltage differential are small. Some linear regulators have small voltage differentials called *low-drop out* regulators—as was described briefly in Chapter 2.

Whatever the application, on-package and on-silicon power conversion appears to becoming popular due to the need to integrate more complex functions into the SoC. As the technology moves forward in time, engineers should expect to see the adoption of such technologies increase, and the power integrity engineer's skills will become increasingly more involved with these types of the power delivery designs.

7.3 ON-DIE DECOUPLING

Decoupling on silicon is becoming an increasingly challenging endeavor as architects and designers continue to push the densities of digital functions within their chip designs. For SoC applications, in particular, the density issues are even more critical as blocks compete for power decoupling while attempting to isolate the rails for better power management. In recent years, there has been renewed interest in creating a high-quality capacitor, such as the MIM cap [8]. For on-die decoupling, the interest here is to gain a more insight into the different decoupling solutions so that PI engineers can understand how it affects the overall PDN system and how to model it.

For complex loads, the capacitance is interspersed among multiple functional blocks, after the primary routing and interconnections have been made. Typically, a given amount of decoupling capacitance is dedicated locally to regions where potential depletion of the charge may occur due to the loading of the functional block. Figure 7.23 gives an example of how capacitance is interspersed laterally (particularly if the capacitance is MOS based) between the units for decoupling. It is not uncommon for areas with high *di*/*dt*'s and/or low noise circuits to have additional capacitance interspersed among them.

The capacitance could also lie directly on top of the functional blocks, but it must be of a different type and be embedded in the layer structure of the silicon. The usage model for the capacitors varies depending on the requirements and need for power decoupling. A number of different types of on-silicon decoupling

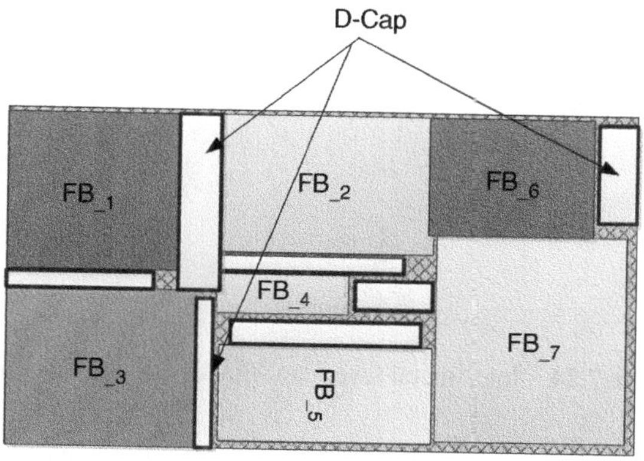

Figure 7.23 Decoupling capacitor interspersed in digital block (functional blocks)

TABLE 7.5 Typical silicon level capacitance and values/area

Capacitor Type	Typical Area Metric
Inter-metal capacitance	$\sim$1–10 fF/μm^2
Transistor (T-cap)	$\sim$5–15 fF/μm^2
MIM cap	$\sim$15–30 fF/μm^2

capacitances are available to chip designers today—depending on process, cost, and constraints. These capacitors are shown in Table 7.5.[7]

Inter-metal capacitance is the capacitance generated using the metal layers and the dielectric between the layers to create the capacitor structure. Typically, this capacitance is small but of high quality. It is used often where only a small amount of capacitance is needed—such as for a filter in an op-amp device—and where it is easily accessible. Normally, this type of capacitance is not used for power decoupling. This is because the values are too small, unless many layers are available to the designer to use. Figure 7.24 shows a simple structure.

As one would expect, the capacitance is different depending on which layer couples to one another. Moreover, there will be a *fringing* capacitance that allows for additional storage in the same basic region. This can be seen in Figure 7.24. The total capacitance is a combination of this fringing capacitance and vertical capacitance. This is written as.

$$C_{Tot} = A \cdot C_v + \sum_{i=1}^{n} P_i \cdot C_{if} \tag{7.12}$$

where A is the area, and P is the subsegment perimeter component relative to the other metal layers. The *fringe* capacitance is computed via a 3D solver. This is

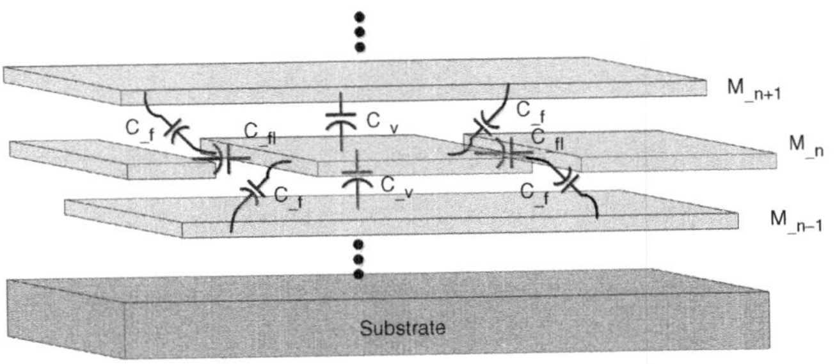

Figure 7.24 Inter-metal layer capacitance structure example

[7]The values in the table vary greatly depending on the process technology, routing constraints, and other design factors. These are only general approximations and designers should ask their manufacturing representative for more accurate data before performing any analysis.

because it is highly dependent on geometry. A simple formula to approximate this (for widely spaced wires) is

$$C_f \cong \epsilon_{ox} \ln\left(1 + \frac{t}{h}\right) \text{ C}/\mu\text{m} \tag{7.13}$$

where ϵ_{ox} is the dielectric constant for the oxide or dielectric, t is the thickness of the metal layer for the main plate, and h is the height above or below the main plate. Notice that the previous equation is a *per-unit-length* relation, given in microns. The lateral capacitance for *closely spaced* wires *laterally* is estimated from

$$C_{fl} \cong \epsilon_{ox} \frac{t}{s} \text{ C}/\mu\text{m} \tag{7.14}$$

where s is the separation between plates, again laterally. Designers compute these directly from the CAD tools in layout, and simple extractions are given to engineers for estimations. For PI engineers, assuming that they have access to general layout rules and have some knowledge of the design, a first-pass estimate for the inter-metal capacitance is easily obtained.

For designs that require larger amounts of local decoupling capacitance, developers use *transistor* level or *T-cap*, utilizing the MOSFET as a transistor. The main limitations are that the connections have to go into the substrate and that it takes up real estate down into the silicon transistor region. The other issue is that the transistor cap is limited to the voltages that they can withstand according to the process, since they cannot typically be made to float as in some metal capacitor structures. Thus, it is common to see one side the capacitor tied to the substrate as its ground. Most mask designers are directed to allocate room after the basic blocks have been placed for some amount of T-cap. Figure 7.25 shows how transistor capacitance is created from the MOSFET structure.

By tying the drain and source together, the MOSFET is used as a capacitor as well. The capacitance of the transistor is found from the following equation:

$$C_{ox} \cong C'_{ox} A \cdot (\text{scale factor})^2 \tag{7.15}$$

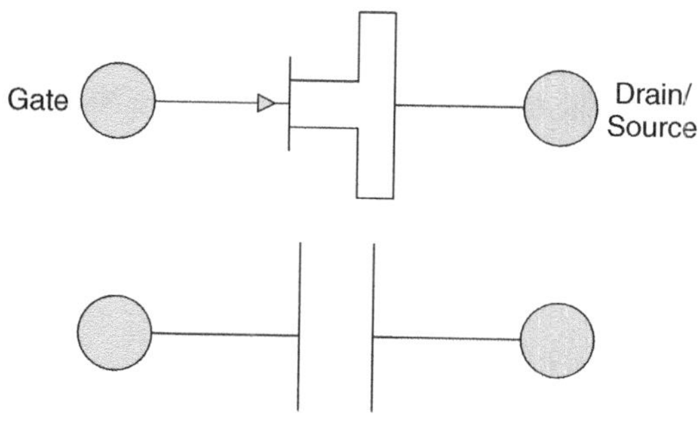

Figure 7.25 N-MOS transistor used as a capacitor

where A is the area, $W \times L$ of the transistor, the scale factor is the process technology, and C'_{ox} is found as,

$$C'_{ox} \cong \frac{\epsilon_{ox}}{t_{ox}} \tag{7.16}$$

The width and length of the transistor are defined as the *effective* length of the transistor and the *drawn* width. This defines the metal geometry for the transistor structure. The number of T-caps used in a system is dependent on many factors—the most important one being the available area left over from the layout and design of the logic region of interest. As shown in Table 7.5, the amount of transistor level capacitance per unit area versus metal capacitance varies with respect to area; however, due to metal-routing constraints and transistor density differences in regions, often the transistor capacitance is more readily available to the designer [2]. After the MD routes the design, there is a larger amount of transistor capacitance as compared with inter-metal capacitance—the final amount per-unit area is highly dependent on the process technology used. There are other methods for creating capacitors using MOSFETs—such as through diffusion capacitance—however, these are more for small filtering applications than for decoupling of the power sources.

The last type of capacitance that designers may utilize (again, depending on the process technology that supports it) is metal-insulator-metal or (MIM) capacitance [10]. A high dielectric constant material is sandwiched between a dedicated metal layer structure in the silicon metal stack, where the material is kept very thin to increase the capacitance density. MIM cap is gaining use in many advanced processes today and in a number of applications, including power decoupling. The advantage is not only higher capacitance density but also a dedicated layer for designers to route and create a high-quality filter.

The next example shows how capacitance is estimated using some of the formulas of this section.

Example 7.3 Determine the capacitance available for a logic unit if the transistor dimensions are 200 × 10 μm, and there are 1000 of these transistor areas available, with the oxide thickness 35 Å; the oxide is SiO_2. Asssume that the process technology is 45 nm.

SOLUTION: A side view of a MOSFET structure connected as a capacitor is shown in Figure 7.26. In the figure the capacitance for the N-MOS structure may be connected so that the positive side of the capacitor is at the gate of the transistor and the negative at the source/drain. When a voltage is applied to the gate, the transistor will forward bias, and a charge will be built up in the path shown. To compute the capacitance, first compute the C'_{ox} term from equation (7.16):

$$C'_{ox} \cong \frac{\epsilon_{ox}}{t_{ox}} = \left(\frac{3.97 \cdot 8.854e^{-12}}{3.5e^{-9}}\right)\left(10^{-12}\right) = 10.04\ \text{fF}/\mu\text{m}^2 \tag{7.17}$$

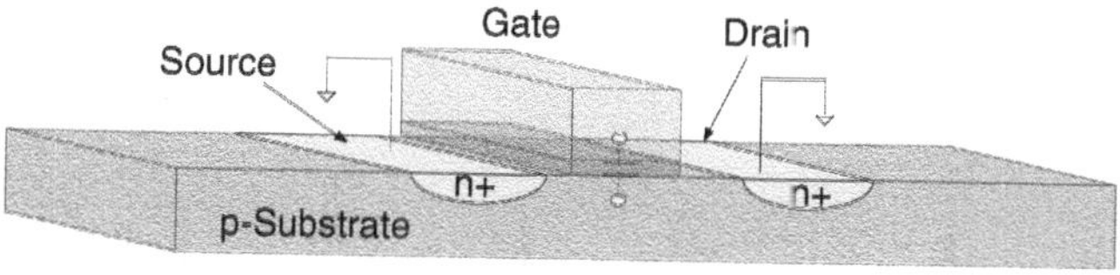

Figure 7.26 N-MOS transistor as a capacitor

The total capacitance is then

$$C_{ox} \cong C'_{ox} A = \left(1.004e^{-14}\right)\ (200 \cdot 10) \cdot (0.045)^2 \cdot 1000 = 40.07\ \text{pF} \qquad (7.18)$$

Notice that the last term is in microns to cancel with the C'_{ox} term. The total area in the substrate that this occupies is

$$0.2\ \text{mm} \cdot (0.01\ \text{mm}) \cdot 1000 = 2\ \text{mm}^2 \qquad (7.19)$$

This area appears to be a large, and for many designs, it is. Nevertheless, note the *size* of the capacitance: 40 pF of capacitance is rather substantial for decoupling a local logic region and is very effective to mitigate HF noise on the local power layer.

The power integrity engineer is interested in the local decoupling for only one reason: to get the proper filtering characteristics for the load filter to develop a meaningful PDN. There was a brief discussion of the capacitance used for load filtering, but there has been no examination of the other lumped parasitics associated with this capacitance. Due to the relatively large number of vias and short traces going down into the silicon, it is common to ignore the *inductive* portion of the interconnect. This is because the inductance from the bumps to the actual capacitance is very small in comparison with that from other parasitics in the PDN. However, the resistance may not be. There are two elements to the resistance that must be considered when gaining insight into these parasitics. First, the interconnect to the capacitors is resistive, and second, the actual plates of the capacitor may also be resistive. As shown in Section 7.1.12, that current may flow through different paths to the load. This is highly dependent on the position of the load relative to the capacitors as well. As seen in Figure 7.27, the lowest impedance charge can flow laterally from, say, a T-cap into the load to support the energy and voltage needed to turn on the transistors.

Thus, the resistance from the filtering is due to a number of interconnect paths, and this typically complicates the analysis. The PI engineer may choose then to consult the designers responsible for the layout to estimate the resistance of the path due to the load. In most cases, trying to get a *first principles* estimation is challenging due to the many elements that go into the distribution network of the load silicon. With the sophistication of the CAD systems today, an MD can perform a simple extraction as well.

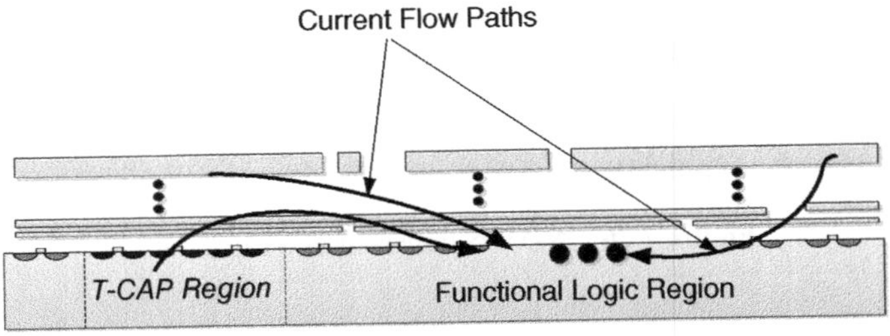

Figure 7.27 Potential current flow paths in silicon cross section during activity

7.4 ADVANCED TOPICS IN POWER ON SILICON

The last topic in this text is on a very new emerging technology in on-die voltage regulation, with major advances in *switching regulators* on silicon. Designers—mainly at universities and in company labs—place the regulators into the silicon and integrate many of the components that used to be external to their silicon. At the time of the writing of this text, there do not appear to be any high-volume production devices out in the market that are completely integrated. However, it is probably only a matter of time before this changes. The main limitations have been the ability to successfully add the *filters* onto the silicon to allow full integration of the switching regulators. Though there has been, in parallel, substantial movement toward the integration of the voltage regulator components over this period as well, the trend toward *full* integration appears to be steadily gaining ground. For the PI engineer, this will eventually have a significant impact on the power distribution path and the PDN—along with the system design that will be required for many of these rails.

Assuming, for the moment, that the power converter is on a distinct die and (possibly) a distinct package—from the load die—there is, at least, a starting point for understanding what the PDN might look like. Though this will be speculative on what this PDN might be, some general assumptions will help in understanding the potential long-term impacts. The goal is to delve into what *kind* of circuits will be generating power from a fully integrated voltage regulator on die.

Earlier in this chapter, there was already some discussion of some of the potential advantages of integration of the power conversion into a single monolithic silicon die. However, now that it is a *switching regulator*, and in particular a *buck regulator*, that is being developed, the subject will require some further examination. First, the characteristics of this voltage regulation scheme should be noted and compared with a platform design. Table 7.6 provides shows a list of key differences between the two approaches.

The main advantages appear to be in cost and area. Voltage regulators close to the load show benefits in capacitor reduction and loop response as was noted in Chapter 5, and earlier in this chapter, because there are fewer parasitics in the PDN. Earlier designs added more and smaller VR blocks onto the die and operated them

TABLE 7.6 Typical characteristics and comparisons between platform and on-die voltage regulation

Characteristics	Platform VR	On-die VR
Number of phases per rail	1–4	2–20+
Number of components	6+ per rail	Typically none
Amount of digital logic	Small	Large
Efficiency (small voltage ratios)	High	High
Cost	Medium	Low*
Input voltage constraints	Less due to older process	More due to closer to logic process
Availability	Common	Not available*
Area	Relatively large	Very small
Power distribution Network components	Large	Typically small

in parallel [12]. For a switching regulator with storage inductors, many inductors would be required to make it operate [11]. One of the advantages of on-die voltage regulation is that smaller, fine-grain power modules can be put into a single silicon die. A disadvantage is that the storage energy in these inductors is small, since the amount of inductance that can be obtained through an inductor on silicon is much less than that of a large inductor on the platform. With the smaller inductor, comes the problem of energy storage, and this necessitates a frequency of operation increase of at least 10-fold from a platform implemented design [13]. The result is that the *effective* efficiency will be lower, but the trade-off is much faster loop response due to the smaller filter and faster switching. Figure 7.28 shows where some developers appear to be heading with the on-die solution.

The set of inductors shown in the figure might have to be spiral, or something close to that, to keep the inductance up. There could be thin-film magnetics on them where the metal inductor routes in and out of the magnetic region in order to increase the inductance [14]. Below these, in silicon, would be both the circuits (bridges, drivers, and controller) and the capacitors for filtering and energy

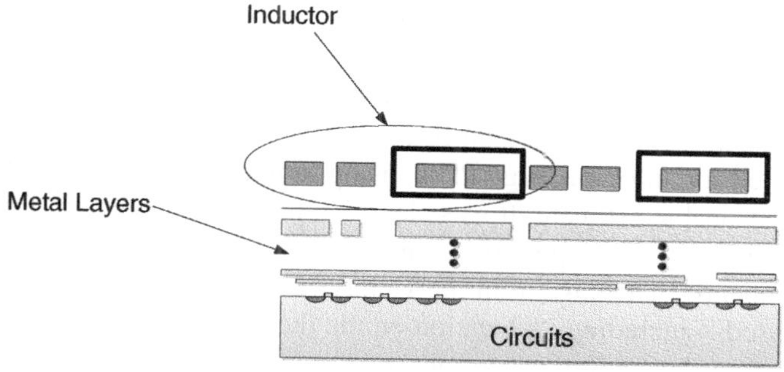

Figure 7.28 Cross section of potential on-die VR structure

storage—along with the inductor and capacitors. A system of this type was presented at APEC in 2010 [12], though there are many possible variants to such a design. From the overall system point of view, the on-die VRs are placed in a separate die relative to the load die and positioned either adjacent to the load die or beneath it on the opposite side of the board and/or package, as was discussed earlier in this chapter.

Interestingly, if there are significant numbers of rails to support it, a single monolithic VR on silicon will have advantages of size and cost—particularly if the *current* load levels to each rail are small. The problem would be how to route out of the die and package the many rails to the load die and other load sources. If the currents turnout to be significant, there would additionally be the matter of increased resistance and inductance of the planes to resolve. Thus, additional filtering would be required.

As designers advance methods of integrating voltage regulator technologies into silicon, they will be continually challenged by routing, noise, costs, area, and a host of other system-level constraints. On-silicon voltage regulation, however, appears to be a number of years out in the future—with respect to HVM, or high-volume manufacturing. Though it is not yet clear where the technology will eventually take hold - it *is* clear that this will have a profound effect on power integrity at the system level, and will present opportunities for designers to create solutions for electronics and computer systems.

7.5 SUMMARY

Silicon power integrity is a relatively new field of study for the power integrity engineer, and there are many challenges ahead. Some challenging issues pertain to the lack of area available for filtering due to the very small space that is available on-die or on-package. As increasingly complex designs, such as SoCs, take foothold in the industry, many more rails will be required on silicon designs. It was shown in Section 7.1 that routing into and out of a device is constrained due to funneling effects from the many different blocks on chip that require a different rail for each block. Additionally, the power distribution is *workload* dependent. Also discussed was the congestion problem between signals and power interconnections, which can occur when current density and routing dominate the PDN. Electromigration issues can arise if not enough copper is provided in the interconnections to the loads from the source up through the silicon. Thus, it is always a good idea to do a quick check to ensure that path resistances are balanced. A simple PDN was then developed for an on-package PDN, using lumped components in the path to arrive at a reasonable estimate. It was clear that the PDN for the platform and for an on-package designs are quite different and estimates for each must be considered independently. Then again, in the vicinty of the load, on-die decoupling has to be addressed—including its location on the die relative to the loads. There are a some different decoupling capacitor structures at the disposal of the engineer, but all are process dependent. As the loads become more complex and segmented, on-die decoupling becomes especially critical for devices such as SoCs. Finally, there was a discussion of on-die switching regulators. For the power integrity engineer, responsible for both the platform and silicon, this is the next generation of PDN technology, as the multi-load and multi-rail system promises new, exciting opportunities, and challenges.

REFERENCES

1. Bohr, M. T. Interconnect scaling-the real limiter to high performance ULSI. *IEEE IEDM*, 1995.
2. Baker, R. J. *CMOS Circuit Design, Layout, and Simulation*, 2nd ed. Wiley, 2005.
3. Gupta, M., Koc, A. T., and Vannithamby, R. Analyzing mobile applications and power consumption on smartphone over LTE network. *IEEE ICEAC*, 2011.
4. Li, W., and Tan, C. M. Black's equation for Today's ULSI interconnect Electromigration reliability—A revisit EDSSC Int. Conf., 2011.
5. Syed, A., Dhandapani, K, Moody, R., NIcholls, L., Kelly, M. Cu Pillar and u-bump electromigration reliability and comparision with high pb, SnPb, and SnAg bumps. IEEE ECTC, 2011.
6. Yang, B., Wang, J., Xu, S., Shen, Z. J. Advanced low-voltage power MOSFET technology for power supply in package applications. *Power Electronics*: *IEEE Trans*, 2013.
7. Araki, T. Integration of power devices-next tasks. *Power Electronics*: *IEEE Trans*, 2013.
8. Klootwijk, J. H., Jinesh, K. B., Dekkers, W., Verhoeven, J. F., Van den Heuvel, oozeboom, F. Ultrahigh capacitance density for multiple ALD-grown MIM capacitor stacks in 3-D silicon. *IEEE Electron Device Letters*, 2008.
9. Kalavathi, S., Kordesch, A. V., Esa, M. Increased capacitance density with metal-insulator-metal—Metal finger capacitor (MIM-MFC). *IEEE ICSE*, 2006.
10. Wu, Y.-H., Ou, W.-Y., Lin, C-C., Wu, J.-R., Wu, M.-L., Chen, L.-L. MIM capacitors with crystalline-TiO2/SiO2 stack featuring high capacitance density and low voltage coefficient. *IEEE Electron Letters*, 2012.
11. Morrow, P. R., Park, C.-M., Koertzen, H. W., DiBene II, J. T., Design and fabrication of on-chip coupled inductorswith magnetic material for voltage regulators. *IEEE Trans. Magnetics*, 2011.
12. DiBene II, J. T., Morrow, P. R., Park, C.-M., Koertzen, H. W., Zou, P., Thenus, F., Li, X., Montgomery, S. W., Stanford, E., Fite, R., Fischer, P. A 400 amp fully integrated silicon voltage regulator with in-die magnetically coupled embedded inductors. IEEE APEC Conf. 2010.
13. DiBene II, J. T. Fine-grain on-die Integrated Magnetics: Breaking through the power/performance barriers. Power SoC Plenary Talk, 2012.
14. O., Mathuna, S. C., Mathuna, S. C. O., O'Donnel, T., Wang, N., Rinne, K. Magnetics on Silicon: An Enabling Technology for Power Supply on Chip. *IEEE Trans. Power Electronics*, 2005.
15. O. Mathuna II, S. C., Mathuna, S. C. O., Wang, N., Kulkarni, S., Roy, S. Power supply on chip (integration of inductors and capacitors with active semiconductors). *IEEE ISPSD*, 2012.

PROBLEMS

7.1 Using Figure 7.3, determine the average power in the SoC over the 200 ms. From the graph, estimate which load dominates, on average, over the other loads. If you were designing a network for this load, what considerations would you make to ensure adequate power integrity?

7.2 Determine the resistance of a rectangular copper that is 89% pure at 105°C if the resistance is 1.4 mΩ. If the copper is 10 μm thick, will there be well-developed skin-effect in this layer when the signal is at the main resonance in Figure 7.19?

7.3 What is the resonance of the PDN in Table 7.4 if L_{PK_VRF} is now 3 pH instead of 300? What is the peak of the resonance?

7.4 Compute the total capacitance of an inter-metal capacitor if the area is 50 × 50 μm, the separation to the adjacent planes are 10 μm, the thickness is 72 Å, and the oxide material is SiO_2.

APPENDIX A

TABLE OF INDUCTANCES FOR COMMONLY USED GEOMETRIES

Inductance computations are the more commonly extracted elements that power integrity engineers use for analytical estimations. In addition to those simple computations shown in Chapter 3, a few more formulas have been added here. There are a number of good references for these closed-form expressions—and many others. Ruehli [1] has described a method to extract *partial inductances*, which allows one to breakdown the total loop inductance of a circuit with partial elements. His method is well worth reviewing because very reasonable approximations may be made for even complex geometries using the well-termed *PEEC* (partial element equivalent circuit) modeling method. Hoer and Love [2] have conveniently given a number of rectangular geometry-based inductances, which are elegant and yield good results for estimation in many problems. Leferink [3–5] also describes a number of planar structures, which are useful references. All of these formulae may be found in Ref. [3]. These formulae represent the *self*-inductances of a few structures only. In most cases, the *mutual* inductance may be ignored for many power geometries. A simple mutual inductance formula for two parallel wires has also been added later for reference. Three excellent additional references are found in [6–8]. Table A.1 summarizes a few of these formulae. Some basic notes follow it

The mutual inductance between two wires of length l may be computed using the following formula [9].

$$M \cong \frac{\mu_0 l}{2\pi}\left[\ln\left(\frac{2l}{s}\right) - 1 + \frac{d}{2l}\right] \tag{A.1}$$

Fundamentals of Power Integrity for Computer Platforms and Systems, First Edition. J. Ted DiBene II.

TABLE A.1 Inductance closed-form expressions

Number	Structure	Equation for L or M
A.1	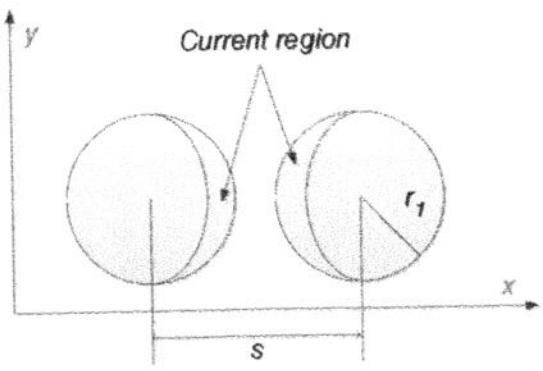	$L \cong \frac{\mu l}{2\pi} \ln\left(\frac{s}{2r} + \sqrt{\left[\left(\frac{d}{2r}\right)^2 - 1\right] - \left(\frac{r}{d}\right)^2}\right)$
A.2	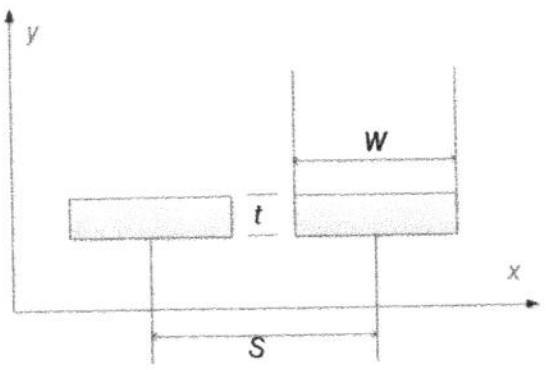	$L \cong \frac{\mu l}{2\pi} \ln\left(\frac{\pi (s - w)}{w + t} + 1\right)$
A.3	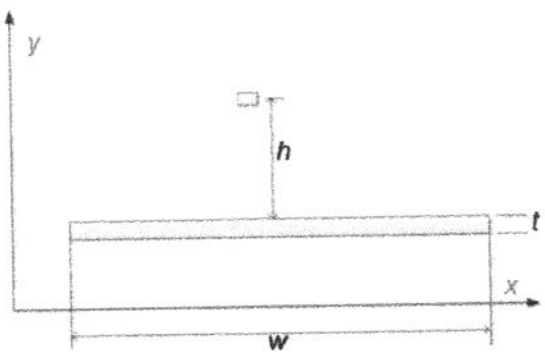	$L \cong \frac{\mu l}{2} \frac{(d - t)}{w}$
A.4	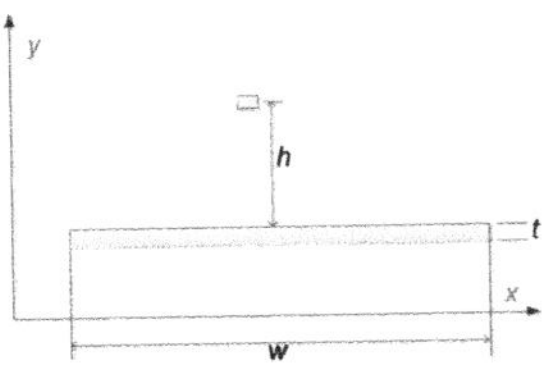	$L \cong \frac{\mu l}{\pi w} \int_t^h \operatorname{atan} \frac{w}{2r} dr$

A.1 – Two wires in close proximity to each other, same diameter. Conditions: I is not uniform, $s \sim r$. When two wires with a voltage applied between them is closer than $5r$ (e.g., the separation between them is less than 5× the radius of the wire), then *proximity effect* can force the charge to accumulate close to one side. Proximity effect is a frequency-dependent phenomenon. When two wires with alternating currents that are equal and opposite to each other are formed in close proximity, current will tend to coalesce in the regions nearest to the gap between the wires. When that happens, the resistance can increase (as well as the inductance). The formula in this section assumes a nonuniform current distribution, which may *approximate* this effect.

A.2 – Two rectangular wires of similar geometry adjacent to one another. Conditions: edge effects ignored. This is a common geometry that may be used for smaller wires where w is greater than t but is not *significantly* greater.

A.3 – Small rectangular wire above a plane. Conditions: $w \gg h$. This formula may be used for structures where a small rectangle is a relatively large distance away from a plane.

A.4 – Small rectangular wire above a plane. Conditions: $w \gg h$, return path is symmetric and directly underneath the wire. Same as A.3.

Again, this is just an approximation. For wires that are separated by fairly large distances ($1 \gg s/l$, which applies to the previous formula as well) where the geometries are planar rather than circular, this formula may also be used.

REFERENCES

1. Ruehli, A. E. Inductance calculations in complex integrated circuit environment. *IBM J. Res. Dev.*, vol. 16, no. 5, 1972.
2. Hoer, C., and Love, C. Exact inductance equations for rectangular conductors with applications to more complicated geometries. *J. Res. Natl. Stand., Sec. C Engineering and Instrumentation*, vol. 69C, no. 2, 1965.
3. Leferink, F. B. J. Inductance calculations; methods and equations. IEEE International Symposium on Electromagnetic Compatibility, Symposium Record, pp. 16–22, 1995.
4. Leferink, F. B. J., and Van Doom, M. J. C. M. Inductance of printed circuit board ground planes. IEEE International Symposium on Electromagnetic Compatibility, Symposium Record, pp. 327–329, 1993.
5. Leferink, F. B. J. Inductance calculations; experimental investigations. IEEE International Symposium on Electromagnetic Compatibility, Symposium Record, pp. 235–240, 1996.
6. Balanis, C. A. *Advanced Engineering Electromagnetics*. Wiley, 1989.
7. Ramo, S., Whinnery, J. R., and Van Duzer, T. *Fields and Waves in Communication Electronics, 3rd ed.*, 1994.
8. Grover, F. W. *Inductance Calculations*. Dover, 1952.
9. Paul, C. R. *Introduction to Electromagnetic Compatibility*. Wiley, 1992.

APPENDIX B

SPHERICAL COORDINATE SYSTEM

The spherical coordinate vector $\boldsymbol{A}$ may be defined as follows:

$$\boldsymbol{A} = A_r \boldsymbol{u}_r + A_\theta \boldsymbol{u}_\theta + A_\phi \boldsymbol{u}_\phi \tag{B.1}$$

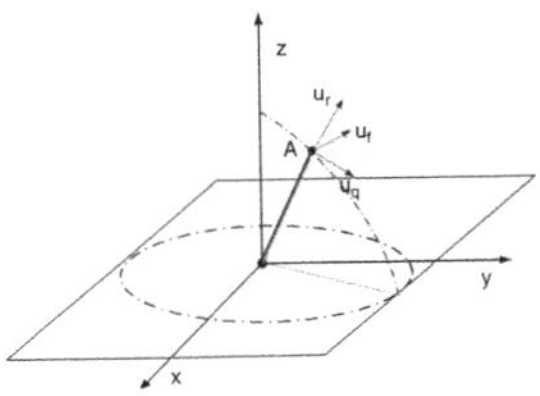

Figure B.1 Spherical coordinate system

where the unit vectors correspond to the unit vectors in Figure B.1.

To convert from rectangular to spherical, or conversely, may be accomplished through the following formulae:

$$x = r\sin\theta\cos\phi \tag{B.2}$$

$$y = r\sin\theta\sin\phi \tag{B.3}$$

$$z = r\cos\theta \tag{B.4}$$

Fundamentals of Power Integrity for Computer Platforms and Systems, First Edition. J. Ted DiBene II.

$$\theta = \tan^{-1}\left(\frac{\sqrt{x^2 + y^2}}{z}\right) \tag{B.5}$$

$$\phi = \tan^{-1}\left(\frac{y}{x}\right) \tag{B.6}$$

A thorough overview of the three different coordinate systems may be found in Paul and Nasar [1].

REFERENCE

1. Paul, C.R., and Nasar, S.A. *Introduction to Electromagnetic Fields, 2nd ed.*, Wiley, 1987.

APPENDIX C

VECTOR IDENTITIES AND FORMULAE

$$\boldsymbol{A} + \boldsymbol{B} = \boldsymbol{B} + \boldsymbol{A} \tag{C.1}$$

$$\boldsymbol{A} \cdot \boldsymbol{B} = \boldsymbol{B} \cdot \boldsymbol{A} \tag{C.2}$$

$$\boldsymbol{A} \times \boldsymbol{B} = -\boldsymbol{B} \times \boldsymbol{A} \tag{C.3}$$

$$(\boldsymbol{A} + \boldsymbol{B}) \cdot \boldsymbol{C} = \boldsymbol{A} \cdot \boldsymbol{C} + \boldsymbol{B} \cdot \boldsymbol{C} \tag{C.4}$$

$$(\boldsymbol{A} + \boldsymbol{B}) \times \boldsymbol{C} = \boldsymbol{A} \times \boldsymbol{C} + \boldsymbol{B} \times \boldsymbol{C} \tag{C.5}$$

$$\boldsymbol{A} \cdot \boldsymbol{B} \times \boldsymbol{C} = \boldsymbol{B} \cdot \boldsymbol{C} \times \boldsymbol{A} \tag{C.6}$$

$$\boldsymbol{A} \times (\boldsymbol{B} \times \boldsymbol{C}) = (\boldsymbol{A} \cdot \boldsymbol{C})\boldsymbol{B} - (\boldsymbol{A} \cdot \boldsymbol{B})\boldsymbol{C} \tag{C.7}$$

$$\nabla \cdot (\nabla \times \boldsymbol{A}) = 0 \tag{C.8}$$

Fundamentals of Power Integrity for Computer Platforms and Systems, First Edition. J. Ted DiBene II.

$$\nabla \times \nabla\psi = 0 \tag{C.9}$$

$$\nabla \cdot (\boldsymbol{A} + \boldsymbol{B}) = \nabla \cdot \boldsymbol{A} + \nabla \cdot \boldsymbol{B} \tag{C.10}$$

$$\nabla \times (\boldsymbol{A} + \boldsymbol{B}) = \nabla \times \boldsymbol{A} + \nabla \times \boldsymbol{B} \tag{C.11}$$

$$\nabla \cdot (\psi \boldsymbol{A}) = \boldsymbol{A} \cdot \nabla\psi + \psi \nabla \cdot \mathbf{A} \tag{C.12}$$

$$\nabla \times (\psi \boldsymbol{A}) = \nabla\psi \times \boldsymbol{A} + \psi \nabla \times \mathbf{A} \tag{C.13}$$

$$\nabla \cdot (\boldsymbol{A} \times \boldsymbol{B}) = \boldsymbol{B} \cdot \nabla \times \boldsymbol{A} + \boldsymbol{A} \cdot \nabla \times \boldsymbol{B} \tag{C.14}$$

$$\nabla \times \nabla \times \boldsymbol{A} = \nabla(\nabla \cdot \boldsymbol{A}) - \nabla^2 \boldsymbol{A} \tag{C.15}$$

INDEX

Fundamentals of Power Integrity for Computer Platforms and Systems, First Edition. J. Ted DiBene II.